HOW TO CREATE A
WILDLIFE POND

KATE BRADBURY

CONTENTS

A POND IN
YOUR GARDEN

INTRODUCTION

What is a pond? A pond is about the best thing you can introduce to your garden. Crudely, it's a body of water with some plants in it. It might be deep or shallow, round or kidney-shaped, formal or wild, edged with wildflowers, decking or paving slabs. It might have a pebble beach, tadpoles and whirligig beetles, or a few pond lilies beneath which newts glide and shelter.

A pond is the centre of your garden. It's life, it's fun. It's where you will be drawn to as you wander around, where you will sit and contemplate the world. It's where you will spot things you've never seen before, marvel at the wonder of metamorphosis and gasp at the cruelty of nature, red in tooth and claw. It's magic, that's what it is. Magic and awe. What is a pond? Why, it's the best thing ever.

WHERE DID PONDS COME FROM?
Ponds have been in existence for millennia. Small bodies of water would form from meandering rivers, natural low-lying depressions in the ground, holes left by fallen trees and beaver activity (which still happens, if we let it). Traditionally, there would be a pond in the corner of every field on farmland for livestock to drink from, while village ponds were essential for providing people with water. These ponds were also stocked with fish for eating. Mill ponds (complete with water wheels) were created to harness the power of water and provide energy for industry. Ornamental ponds became grand focal

Pond in summer
Lush with growth, a pond in summer is full of life.

points in stately gardens across the world, which led to our parks and gardens featuring smaller versions of them, for a little piece of glamour outside our back door.

With every pond, big or small, formal or wild, wildlife has seized an opportunity. Amphibians may have evolved in footprint ponds made by dinosaurs and woolly mammoths. Later they would have ducked beneath the surface when a cow or horse dipped its mouth into a dew pond for a drink or when villagers jumped in for a wash. Many invertebrates have also evolved with ponds, either relying on them their whole lives or using them during part of their lifecycle. Ponds are essential for virtually all forms of life, from amphibians, fish and insects such as dragonflies and water beetles, to mammals and birds and even bees, for drinking. A pond is all things to all things.

A POND FOR ALL SEASONS

With a pond in your garden, you get to sit back and enjoy watching all the life that

the pond brings in. At home, I stumble out each morning, while the kettle boils for tea, and look into the water. In spring I might see mating frogs and frogspawn, backswimmers and pond skaters, the first toad of the year. In summer I see beetles zipping around the pond edge and dragonflies dipping their abdomens into the water to lay eggs. Around the pond, masses of bees, flies and other insects gather on the flowers growing at the edges, while baby amphibians jump or crawl around my bare feet. In autumn I

Icy beauty
A frozen pond offers little for those species above the surface, which are unable to use it for drinking and bathing. However, frogs, beetles and other pond life survive sheltered in the mud at the bottom.

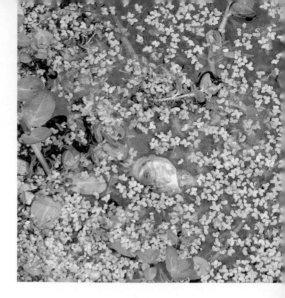

look for hedgehog trails and droppings,
or bird footprints in the mud. In winter I
might see ice on the surface, the morning
sunlight reflecting back on frosted tufts
of grass at the edge. I get lost in my pond
– I end up drinking cold tea nearly every
day. *What was that? What's next? Who are
you?* Questions I ask all the time, thanks
to a little body of water with some plants
in it, outside my back door.

WILDLIFE BENEFITS OF PONDS

Garden ponds support a host of
wildlife, from the fish, amphibians and
invertebrates that rely on them to breed,
to the mammals and birds that need
them to drink and bathe. A pond will be
the most biodiverse part of your garden,
often within weeks of you digging it.

It's difficult to overestimate the
importance of ponds for wildlife. In the
UK, ponds have steadily declined in both
number and quality in the last century.
Some 13 per cent less wildlife are using
the remaining ponds than in the 1970s.
While ponds are still widespread in much
of Europe, some countries, such as the

Netherlands and Switzerland, have lost
90 per cent of their ponds in recent years.
Dew ponds have been replaced by cattle
troughs, natural ponds have been filled
or left to silt over and garden ponds
have been removed in favour of low-
maintenance "outdoor rooms" and
smaller spaces.

A network of ponds can act as
stepping stones, between which a huge
variety of species can travel. Fewer ponds
mean some species can't travel as easily,
as they need to be near a source of water.
This might cut off populations, leading to
them becoming inbred and less resilient
to disease and natural disasters. Cut-off

Drinking spot
A well-placed stone provides
the perfect spot for this wren
to stop for a drink.

populations will also be less able to adapt to the effects of climate change or other changes to habitat. And so, gradually, fewer ponds leads to less wildlife. It's as simple, and sad, as that.

PONDS AND PEOPLE

We benefit from ponds, too. Studies have shown that being in, on or near water improves our mental health. Our blood pressure lowers and we feel calmer. Having even a small pond in your garden might make you feel better. And you'll feel good about throwing a lifeline to wildlife that relies on water to exist, too.

> Ponds are stepping stones. Without them wildlife populations can be cut off.

I love ponds. I can't fully explain why but I know that I'll never live without one again. We didn't have a pond in the garden I grew up in but there was one next door and I was always peering through the fence for a glimpse of it.

Then, aged six, I "accidentally" fell into a pond in a garden centre and my mum firmly refused any further requests to let me dig one until I was 24 (and had long since moved out). I dug two, for good measure. Now I have a pond in my back garden, a small container pond in the front garden, and another pond on my allotment. I'm contemplating a fourth, you know.

I sit by them all and look into them. I breakfast and lunch with them, lie in the sun by them, rest while gardening next to them. I work from home and set myself little deadlines, after which

Ponds are a complete world of weird and wonderful, with layers of complex life. They make me happy.

I'll take five minutes by the pond. Sometimes I'll see something beautiful, like a baby frog or a bathing bird, other times it will be a backswimmer eating a still-live fly – ponds can be pretty gruesome, too. But that's what makes them alive! That's what makes them fascinating! I think that's exactly what I love about them. They're a complete world of weird and wonderful, with layers of complex life. They make me happy. I think they'll make you happy, too.

Grassy margins
Long grass at the pond edge provides shelter for a huge range of species.

DESIGNING YOUR POND

POND PLANNING

I love the planning stage of pond-building: working out where to put it, its shape, what extra wildlife habitats I'll create in and around it. Crucially, I like to plan where I'll spend time watching all the wildlife that uses it. Choosing where to sit and enjoy your pond is an important consideration, and shouldn't be neglected.

My kidney-shaped pond sits two-thirds of the way up the garden on the sunny side. It has two beach areas (see p.24) – one on each end – so I have a clear view across the pond and can watch birds bathing from my kitchen. It's lined with mud so you can't tell where the garden ends and the pond begins, and has a dragonfly perch, mainly used by house sparrows. Around the pond I've planted low-growing bird's foot trefoil and marginals such as water forget-me-not and brooklime. In the pond are native oxygenators plus other marginals with wonderful names like arrowhead and bogbean. Beyond is my small wildflower meadow, through which hedgehogs have created "desire paths" leading to the water's edge.

I enjoy viewing the pond from several angles – the step in front of the shed, where I sit and eat my breakfast, and from the other end, where I lie and gaze into the water. Recently I put in a bench, where I sit by the pond and stare at it for hours.

The following pages cover some aspects to consider while planning. More practical projects are covered in the next chapter (pp.30–67).

Pond near the house
Consider at the planning stage how you want to enjoy your pond. It's easier to spend time by a pond close to the house.

How big your pond is depends on how much of a feature you want it to be in your garden and how much wildlife you aim to attract. In general, the larger the pond, the more wildlife it supports, but even a container pond attracts a range of species.

DEPTH AND DIAMETER

Ponds to attract wildlife don't need to be as deep or as wide as you might think; many ponds that form naturally are small and shallow. Whether formal or informal in appearance, a pond with a diameter of just 2m (6½ft) and a maximum depth of 30cm (1ft) will be perfect for a range of species including invertebrates and some amphibians. Bear in mind that you may need to top it up in summer (see p.178) and stop it freezing in winter (see p.174).

That said, if you have a huge space and want to make a feature of your pond, your garden is your oyster. A large pond can include anything from fountains to a viewing platform. You can have it all – deep areas for fish and wildfowl, shallow pools for invertebrates and aquatic larvae to gather, and the widest range of plant life possible.

If you have space, aim for a pond with a range of depths, including shelves and "shores" (shallows). This creates the greatest range of planting opportunities, where different invertebrates can feed, seek shelter and lay eggs. Ponds are enormously complex habitats. Make yours as complex as possible, and you'll be rewarded with masses of wildlife.

USING CONTAINERS

Container ponds (see pp.52–59) can be perfect for small gardens, patios and even balconies. An old Belfast sink or Victorian baby bath takes less than 1sq m (11sq ft) of garden space, and provides drinking water for mammals and birds,

Patio pond
Even in the smallest space you can enjoy beautiful pond plants and some pond wildlife.

along with breeding opportunities for some amphibians and invertebrates. The best thing about container ponds is that they can be temporary. If you need to move you can take your pond with you.

LINING YOUR POND

In this book I focus on ponds made with flexible liners. You can buy pre-formed, plastic liners, which are typically designed for keeping fish. As they tend to be deep with steep sides and no shallows, I've not included them here. Other types of liner, such as cement and clay, are suitable for large ponds only.

HOW MUCH LINER WILL I NEED?

To work out how much liner (and underlay) you need, multiply both the maximum width and length of your pond by twice the maximum depth. So, if your pond has a maximum length of 4m (13ft), a maximum width of 2m (6½ft), and a maximum depth of 60cm (2ft), you will need:

Length: 4m x 1.2m = 4.8m

Width: 2m x 1.2m = 2.4m

Round up your quantities to be on the safe side; in this instance, you would order 5m x 3m of liner. You can also use an online pond liner calculator.

Choosing the right spot
This pond benefits from some shade from the tree, but you may need to clear leaves from the surface in autumn.

SITING YOUR POND

Before you pick up a spade, think carefully about where you dig your pond. All being well, you'll be living with your pond for years, so you want a spot where it will look good, and will stay in top condition throughout the seasons.

SUN OR SHADE?
A sunny site may appear to be the best option for a wildlife pond, as the warmth and light will enable algae to grow faster, providing more food for algae-eaters like tadpoles, water fleas and pond snails. However, if algae grows too quickly, it can cause problems (see pp.172–173). Too much sun can also cause water levels to fall significantly in summer, especially in a small or shallow pond. An entirely

shaded pond, on the other hand, will typically have less wildlife.

Ideally your pond should receive both sun and shade. In my west-facing garden the pond is in full sun for a few hours of the morning. The rest of the time some of it is in shade. This helps deter algal blooms and stops the water level dropping so quickly, while still providing habitats for sun-loving wildlife.

CONSIDERING TREES

Try to avoid digging your pond beneath deciduous trees. While they may cast much-needed shade in summer, you'll forever be raking leaves off the surface in autumn. A few leaves can be good for ponds, but in large numbers they clog up the water and cause problems in winter if the pond freezes over. Tree roots can also be a problem: they make it harder to excavate the soil, and can lead to a misshapen pond and liner punctures.

PERSONAL PREFERENCES

Think about how your pond will impact on other uses of your garden. Are you happy to give up part of your ornamental border for a pond? Will you still be able to play football/croquet/cricket or have picnics on your lawn, and if not, will this worry you? Think about how the pond will look from your back door as you view the garden. Take into account seating, where

you can make the most of this new habitat you're creating.

Finally, consider children and grandchildren, even if they don't exist yet. Is there an area you can put the pond while children play safely in another area? If not, can you leave space to erect a small fence around your pond when you need to? Ponds offer countless hours of fun for children, but care should always be taken with young children at the water's edge.

POND SHAPE

The shape of your pond can determine how attractive it is to wildlife. Kidney-shaped ponds typically provide the greatest wildlife opportunities, but it's important that your pond works for you, too. Consider the style of your garden and what would work best in your space.

As a general rule, if you have more wavy edges to your pond, you will create extra shallow pockets, attracting more wildlife. These shallows are microhabitats that offer variety and enable more specialist species to thrive. A sqaure or rectangular pond will not be able to create these microhabitats, even with a shallow area.

The shape of your pond will be more noticeable with formal edging such as decking or paving stones (see pp.42–51) than if you leave the edges "wild" and lined with plants, which will spill over the edges and into the water (see pp.32–41).

SQUARE OR RECTANGULAR

These ponds tend to suit a formal design, and are typically edged with paving slabs or decking. Sometimes they're raised off the ground so you can sit on the edges. They tend to have fewer opportunities for shelves and shallows, and you may need to add an access ladder for mammals and amphibians to climb in and out.

CIRCULAR

Being symmetrical, circular ponds also lend themselves to formal edging but they make excellent wildlife ponds, too. Add a beach area, dead wood and some shallow edges to make it perfect for invertebrates. Consider adding wavy edges as part of the overall circle design to make little pockets of microhabitat, where invertebrates can gather.

IRREGULAR OR KIDNEY-SHAPED

This is the best design for a wildlife pond, where you can really go to town on habitat creation. It has plenty of scope for shelving and shallows and you can add beach areas, soft edges, and features such as logs and other forms of dead wood, plus large stones, to further enrich the habitat for as many species as possible.

Square or rectangular

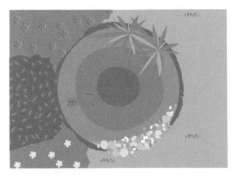

Circular

Irregular or kidney-shaped

A pond is a fantastic habitat in its own right, but adding elements in and around your pond can make it even more welcoming to wildlife. These increase or enhance the habitats available, connecting your pond to the wider garden environment.

WATER FOUNTAIN OR PUMP

It's a common misconception that ponds need a fountain and filtration system to oxygenate the water. This is true only for fish ponds, where the waste produced by the fish, and excess food, can lead to low oxygen and high nitrate levels. However, a fountain or small cascade can make a pleasing addition to a wildlife pond, offering gentle sounds and helping to oxygenate the water. A fountain can also be a grand focal point in a formal pond. You will need to ask a professional to lay pipes and electricity cables while the pond is being built. At the other end of the scale, you can simply place a cheap solar-powered fountain on the surface.

BEACH AREA

Adding a pebble beach to a sunny area of your pond will create a tadpole nursery that will provide them with food as well as shelter from predators. Tadpoles rest in the shallows as temperatures are warmer here than in other parts of the pond, and their main food source, algae, develops faster. Pebbles provide the perfect habitat for algae to grow on and they also provide nooks and crannies in which tadpoles can hide from predators. Beach areas tend to slow down the growth of marginal plants, creating a clear, shallow area for birds to bathe.

DRAGONFLY PERCH

No wildlife pond is complete without a dragonfly perch, such as an arching branch pushed into the soil beside the pond. Dragonflies may perch on the branch to hunt and defend their territory against competing males, which they will then try to see off. Save a good-sized branch from tree prunings to use as the perch. Birds may use it, too, so push it into the ground securely (taking care not to puncture the liner) so that it will take their weight. If you build a log pile (see below) in a sunny spot near your pond you may spot dragonflies perching on this too.

LOG PILE

Stack some logs near your pond and a wealth of wildlife will use them, from beetles and other invertebrates in the nooks and crannies formed by gaps and peeling bark, to amphibians in damp spots within and beneath the pile. If your

Water fountain or pump

Beach area

Tadpoles congregate in the shallows, where temperatures are warmer.

Dragonfly perch

Log pile

Reptile refuge

Compost heap

Add logs to your pond – some invertebrates will breed in the wet, rotting wood.

Stone pile

Wildflower area

log pile is in a sunny area, you may even spot insects and reptiles basking on the sunniest logs to warm up. Consider making a sunny log pile as well as a shady one, as they will likely attract different species. Log piles offer useful protection from cats as amphibians, reptiles, small mammals and even birds can jump into them for safety. You can also add logs to the pond itself. Some invertebrates will breed in the wet, rotting wood.

COMPOST HEAP

As well as being a useful addition to a productive garden, an open compost heap provides hundreds of opportunities for amphibians and other animals, which will rest here, hunt invertebrates and may even hibernate over winter. Some reptiles are drawn to the warmth of compost heaps and – if you're lucky – may breed in them. Always be mindful of amphibians when you dig out or turn your compost heap; it's easy to harm them accidentally.

STONE PILE

A pile of stones will provide damp nooks and crannies for amphibians to shelter from predators, while they are also hunting for invertebrate food. A small pile is sufficient. You could also use large stones to edge your pond, or even build a dry stone wall to provide similar habitats in your garden.

REPTILE REFUGE

Laying pieces of slate, corrugated iron, or even carpet in sunny spots around your pond will create basking areas for cold-blooded reptiles, such as slow worms and grass snakes, which need to warm up under cover. You'll also find invertebrates here, which reptiles and other animals eat.

WILDFLOWER AREA

Wildflowers around your pond will draw pollinating insects, making the area even more biodiverse. This benefits bats, which already have a feast of mosquitoes and midges from the water itself.

Simply scatter wildlflower seed on bare soil in spring or autumn, following the packet instructions. You can also buy wildflower turf, made up of grasses and wildflowers, to lay beside the pond. Lay turf in autumn or spring. Prepare the ground by removing perennial weeds and raking the soil to level it. Lay the turfs close together as there will be some shrinkage, which can cause gaps. Water well and continue to do so every few days, unless heavy rain is forecast. Do not walk on them while they establish. Alternatively, let grass around the pond grow long and see which wildflowers turn up. You may be surprised by how many species are willing to seed into your garden naturally, if you let them.

CREATING
YOUR POND

RIGHT POND, RIGHT PLACE

My three ponds are all different. There's the kidney-shaped one in the back garden, a shallow circular pond on the allotment, and my front garden container pond, which is an old Belfast sink with some marginals and oxygenating plants in it, plus an access ladder for frogs.

Each of my ponds is tailored to suit its surroundings. I chose a deeper pond for the back garden because I wanted to attract toads, which tend to prefer spawning in larger ponds, often with fish (sorry toads, there are no fish). Both frogs and toads seem to like it, plus hedgehogs, birds and a wide range of invertebrates.

The allotment pond is smaller and shallow, partly because I have less space but also because I wanted to make the ultimate frog and invertebrate pond – a good-sized shallow area where tadpoles and aquatic insects and their larvae can shelter. This was a hit with frogs, who found it in the pond's first summer and started spawning there by its second spring. Now several pairs spawn in it – I counted 12 blobs of spawn last spring.

In my small front garden I wanted to make sure amphibians, hedgehogs, foxes and birds had access to water if they couldn't reach the back garden. I also regularly find mosquito and midge larvae in there, which attract – and are eaten by – dragonfly nymphs. Some pond snails must have sneaked in via the pond plants. I sit sometimes with a cup of tea in the front garden, next to the pond, and watch the snails glide below the surface.

Large garden pond
With its huge variety of plants, this pond attracts plenty of wildlife.

WILDLIFE RATING 10/10   **TIME TO BUILD** 1–2 days   **EASE OF MAINTENANCE** 5/10

Project 1

WILDLIFE POND

Wildlife ponds are the most natural of all ponds. Designed to resemble pools and ditches you would find in the wild, the best ones are well planted, with a host of marginals around the edge and have no visible liner. They form a bustling wildlife refuge, attracting amphibians and reptiles, birds and mammals, along with a vast array of weird and wonderful invertebrates.

YOU WILL NEED

Length of hose or string

Spade

Long piece of wood and spirit level

Two large pieces of fleece underlay, enough to fully line the pond twice (see p.21)

One large piece of butyl pond liner, enough to fully line the pond once (see p.21)

Bucket and trowel

Stones, logs, and long branch

Kitchen scissors

Plants (see p.35)

Wildflower turf or other edging plants

Wildlife pond gallery
Grassy margins (near left) create a natural look and enable young frogs and breeding toads to hide from predators. Stones at the edge and underwater (far left, top and bottom) provide hiding places for amphibians.

A wildlife pond typically has gently sloping sides so mammals can get in and out easily, shallows so birds can bathe and insects breed, and native plants to mimic natural ponds (see p.176). Rather than having a hard surface such as paving around the edge, it is edged with grass, low-growing herbaceous plants, or a wildflower mini-meadow instead (see p.27), to provide amphibians and other animals with shelter. Water levels can fluctuate dramatically as plants near the edge can draw water from the pond.

This design uses a clever trick to make the pond look completely wild. A second layer of underlay is placed over the pond liner and covered with a thick layer of nutrient-poor subsoil from the bottom of the pond hole. This hides the pond liner, and allows plants to root naturally. While

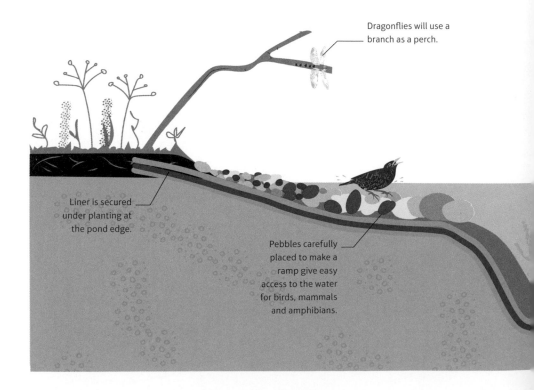

Dragonflies will use a branch as a perch.

Liner is secured under planting at the pond edge.

Pebbles carefully placed to make a ramp give easy access to the water for birds, mammals and amphibians.

Plants root naturally in the pond, to beautiful effect,

the water may look muddy for a few months (which is normal), the roots will knit together to create a stable, long-lasting, natural-looking pond.

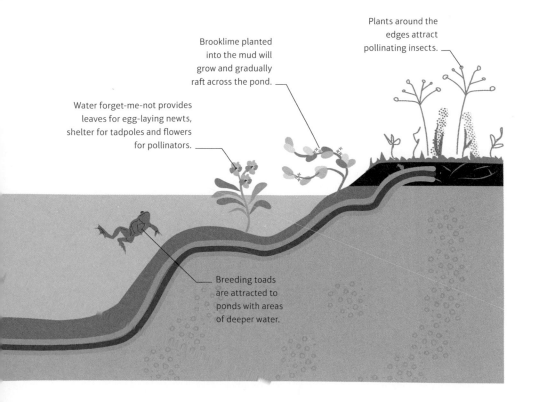

Plants around the edges attract pollinating insects.

Brooklime planted into the mud will grow and gradually raft across the pond.

Water forget-me-not provides leaves for egg-laying newts, shelter for tadpoles and flowers for pollinators.

Breeding toads are attracted to ponds with areas of deeper water.

WILDLIFE POND STEP-BY-STEP

STEP 1

Choose the site of your pond carefully. Ideally, it should get sun in spring when amphibians are spawning, but partial shade in summer so that it doesn't dry out. Use a length of hose or string to lay out your pond in your chosen shape and size.

STEP 2

Starting in the middle, dig your pond. Aim for a depth of 60cm (24in) in the centre, graduating to just 10–30cm (4–12in) around the edge, with plenty of gently sloping shallows. The shallows are the most important areas of a wildlife pond, where birds will bathe, mammals will drink, and amphibians and invertebrates will lay their eggs.

STEP 3

As you work, check that the sides of your pond remain level with each other by laying a piece of wood and spirit level across the centre, from edge to edge. As you dig down, you'll reach subsoil, which is paler in colour than the topsoil. Once you're happy with the depths and shape of your pond, carefully remove stones and any sharp items from the soil so that they don't poke through the underlay that you are about to lay down and tear the liner.

Use a length of string to mark out possible positions and shapes for your pond.

Keep adjusting the shape of your pond until you are happy with it.

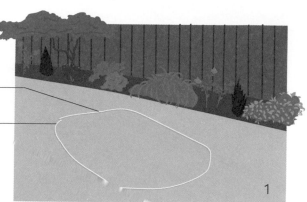

1

Dig out your gently sloping layers using a spade.

You don't need to make flat "shelves" for your plants as you will be taking them out of their baskets to plant.

2

Topsoil

Keep the subsoil that you dig out to one side as you will need to use this later.

Check regularly that the sides are level.

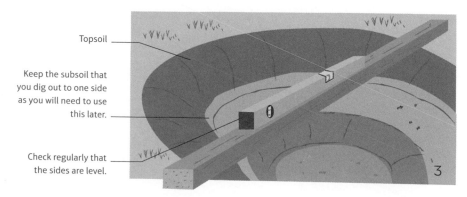

3

STEP 4

Unroll the underlay over the pond so there's about 50cm (20in) of excess at either side. Take off your shoes and carefully climb inside to press the underlay down and smooth it out, folding where necessary. Don't cut any excess away just yet. Once the underlay is in place, repeat with the butyl liner, smoothing it over the underlay. Try to keep folds to a minimum. Climb out of the pond and check that the liner fits as snugly as possible and covers the underlay completely.

STEP 5

Now it's time to lay a second layer of underlay over the liner, again carefully following the technique described above to obtain a neat, smooth surface. Weigh all the layers down with stones to keep them in place.

STEP 6

Fill your bucket with the subsoil you set aside, adding water to make it more pliable if necessary. Starting in the deepest part of the pond, use the trowel to place a layer of around 10cm (4in) of subsoil over the underlay. Keep doing this until you've completely covered the underlay with subsoil. This is a messy but satisfying job.

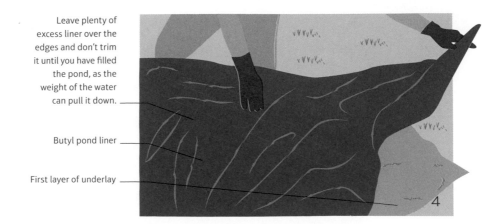

Leave plenty of excess liner over the edges and don't trim it until you have filled the pond, as the weight of the water can pull it down.

Butyl pond liner

First layer of underlay

4

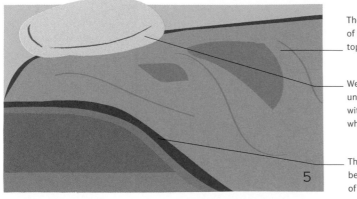

The second layer of underlay sits on top of the liner.

Weigh down the underlay and liner with smooth stones while you work.

The liner is sandwiched between two layers of underlay.

5

Keep the stones in place until the subsoil reaches the top of the liner.

Firm the subsoil as you add it, using the back of the trowel.

Subsoil contains fewer nutrients than topsoil, so will discourage algae from growing strongly.

6

STEP 7

Now's the time to dress the pond. Add stones to one area to make a beach, pressing them firmly into the mud. You can add logs and other pieces of wood, which wildlife can use to rest on or hide among. If you have a long stick to make a dragonfly perch, push this into the soil around the pond (taking care not to puncture the liner), so it hangs over the water.

STEP 8

Remove marginal plants from their baskets and divide the rootballs into separate portions. Plant these portions directly into the mud all the way around the edge of the pond, paying attention to their ideal planting depths (see pp.72–73). Pack the roots down into the mud. The plants may not look like much but they'll quickly grow.

STEP 9

Fill the pond, ideally with rain water, but tap water is fine at this stage (see p.178). Once the pond is full, cut back the excess liner and underlay from the edge. Trim back the top layer of underlay further than the pond liner as it can sometimes act as a wick, drawing water from the pond to the top. Finally, add your chosen edging, be it plants, seeds or turf. If using wildflower or lawn turf (see p.27), gently roll this out around the edge. Firm it with your feet and cut away excess with a bread knife or similar.

A stony beach area is useful in a wildlife pond. Use a mixture of shapes and sizes.

Crevices between the stones are valuable shelter for water invertebrates.

7

Smooth the mud down again after planting so that the underlay is not exposed.

Firm the plants into the pond mud so the roots are fully covered.

8

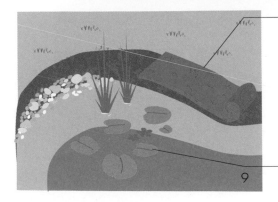

Wildflower turf creates a meadow habitat beside the pond, which animals will use for shelter. It grows tall in summer so may not be appropriate all the way round.

The pond water may look muddy initially but will settle as plants start to establish.

9

Project 2

FORMAL POND

Also known as an ornamental pond, a formal pond has a smart, symmetrical design – usually square or rectangular – and is often a focal point. It canbe raised off the ground or, as in this project, sunken and edged with slabs. While there are typically fewer plants in a formal pond, it can still offer a home to wildlife if access is easy and a shallow area is included in the design.

YOU WILL NEED

Tape measure and length of hose or string

Small sticks to mark corners

Spade

Long piece of wood and spirit level

One large piece of fleece underlay, enough to fully line the pond (see p.21)

One large piece of butyl pond liner, enough to fully line the pond (see p.21)

Levelling pegs

Kitchen scissors

Cement powder and sand for concrete and mortar

Builder's trowel

Edging slabs

Plants (see p.45)

Formal pond gallery A bridge (near left) over two square ponds creates hiding places for animals below. Add a fountain or water spout (far left, top) for a focal point. A circular design (far left, bottom) makes the perfect formal display.

Many gardeners think of formal ponds as deeper than wildlife ponds, but they need only be 60cm (24in) deep at most. Unlike many wildlife ponds, formal ponds may have a pump and filtration system to keep the water clear – typically in the form of a fountain or statue with a water spout (see p.24). To make a formal pond safe for wildlife, it's important to include a feature that lets animals enter and exit safely; a large stone or grooved wooden ramp placed near the edge may be all that's needed. Formal ponds edged with paving slabs can be tricky for amphibians in very hot summers. The slabs can heat

Site the pond in partial shade to make it easier for amphibians in hot sun.

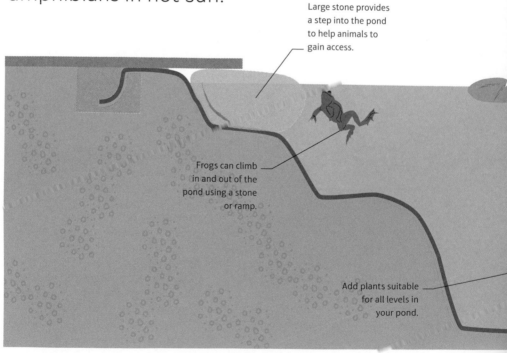

Large stone provides a step into the pond to help animals to gain access.

Frogs can climb in and out of the pond using a stone or ramp.

Add plants suitable for all levels in your pond.

up, causing amphibians to become stuck when they exit the pond, so it's a good idea to site the pond in partial shade.

This design is for a sunken formal pond with a shallow area and additional access so the pond is safe for wildlife. Instructions for fitting a water pump vary, so follow guidance on the packaging if you plan to add one.

SUGGESTED PLANTS

MARGINALS: **Yellow-flag iris**
(*Iris pseudacorus*)
OXYGENATORS: **Hornwort**
(*Ceratophyllum demersum*)
FLOATING PLANTS: **Water lily**
(*Nymphaea*)

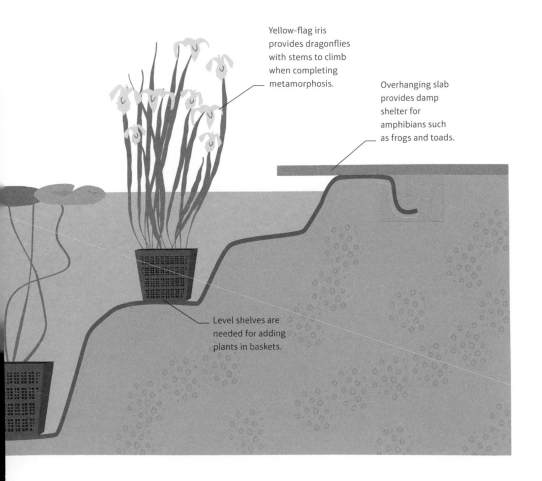

Yellow-flag iris provides dragonflies with stems to climb when completing metamorphosis.

Overhanging slab provides damp shelter for amphibians such as frogs and toads.

Level shelves are needed for adding plants in baskets.

FORMAL POND STEP-BY-STEP

STEP 1

If you plan to add a pump or fountain, ask a professional to visit before you start your pond to discuss laying pipes and electricity. Using a tape measure and string, mark out your pond. Measure the width of your edging slabs and make a second outline to accommodate these. Within the pond area, mark out your depths – three tiers of 60cm (24in), 30cm (12in), and 15cm (6in) from your pond edge will work well.

STEP 2

Dig the outer edging area to 10cm (4in) deep. This will become the trench where your concrete and mortar mix will sit for the edging slabs. Make sure that it's level all the way around, using levelling pegs. Knock in a master peg then use a plank of wood and a spirit level to align the other pegs.

STEP 3

Leaving a 10cm (4in) gap from the inside edge of the trench, start digging your pond from the outside in using the depth markings you have already made as your guide. For stability, make the sides of your pond slope slightly inwards. Make sure each tier is wide enough to accommodate the planting baskets of your chosen pond plants. Check that the top edge of the pond and tiers are level, using a wooden plank and a spirit level.

Take time to mark out the shape and depths of your pond, as this will make it easier to dig and edge later.

Mark the pond and the trench before starting to dig.

Mark each corner with a stick pushed into the soil.

1

The trench will be filled with concrete to form a base for the edging slabs.

Use a spirit level on a plank on levelling pegs to ensure the depth is even all the way round.

2

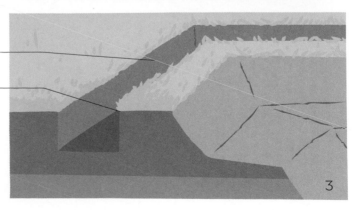

The trench extends around the pond edge.

The 10cm (4in) gap between the trench and the pond forms a ridge. The pond liner will lay over the ridge into the trench, beneath the paving slabs.

3

STEP 4

Remove any protruding stones and other sharp items such as glass from the soil, then lay the underlay over the pond and trench. Remove your shoes and carefully climb into the pond to smooth the underlay out, folding it over where necessary. Repeat with the butyl liner, weighing it down with smooth stones if you need to. Don't cut away any excess underlay or liner – the edges should at least sit comfortably in the trench around the pond.

STEP 5

Part-fill the pond with water. As it fills, gently tug at the edge of the liner to remove any small creases. Take care to ensure that the corner folds are tight. Stop filling the pond when it's about two-thirds full, so that you'll be able to add the edging slabs without water getting in the way. Trim away any excess liner and underlay that extends beyond the trench.

STEP 6

Mix the concrete, following the manufacturer's instructions, then pour it into the trench around the edge of the pond, covering the liner and underlay. Make sure that the concrete is level with the top of the ridge all the way around, then leave to set for at least 24 hours.

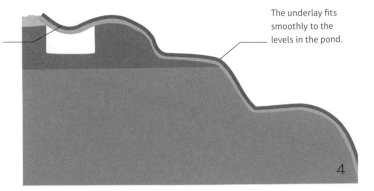

Make sure the liner sits tightly in the pond and extends over the ridge and into the trench.

The underlay fits smoothly to the levels in the pond.

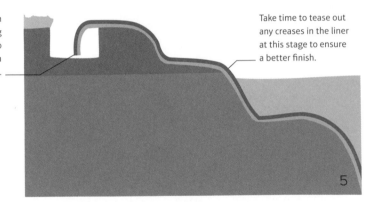

Once the water is in and the liner is fitting well, trim the liner to sit comfortably in the trench.

Take time to tease out any creases in the liner at this stage to ensure a better finish.

Follow the manufacturer's instructions to ensure your concrete mix is the right consistency.

Avoid walking near your pond for 24 hours while the concrete is setting.

STEP 7

Once the concrete has fully set, prepare the mortar by combining four parts sharp sand to one part cement. Apply a level layer over the concrete and lay the slabs onto it, allowing 10mm ($\frac{1}{2}$in) between the slabs for the mortar joints. Ensure that they overhang the pond by 5cm (2in) so you can't see the pond liner. Firm them in place. With more mortar, fill in the joints between the slabs using a builder's trowel.

STEP 8

Once you have pointed between the slabs, you can leave the joint between the slabs and the liner unpointed – this cool dark spot is a perfect habitat for amphibians to rest. Or, to give the pond a neater finish, seal the joint with mortar. To do this, empty the pond using a bucket, then climb inside (without disturbing the setting slabs), and use a trowel to point the joint. Clean the liner well, as mortar is poisonous to plants and wildlife. Let the mortar dry for 24 hours.

STEP 9

Refill the pond then plant up your pond plants in baskets and place them on the most appropriate tier (see pp.72–73). Finally, add a large stone to the shallowest pond tier, which will enable mammals to move in and out of the pond safely.

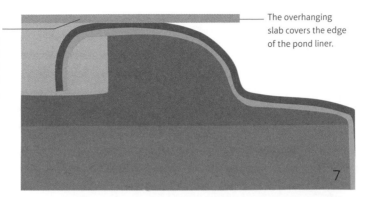

The more accurate and even the 10mm ($^1/_2$in) gaps between the paving slabs, the better the overall look of your pond.

The overhanging slab covers the edge of the pond liner.

7

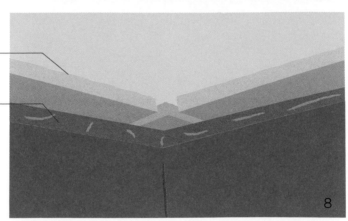

Paving slabs overhang the pond edge.

Mortar forms a layer to secure the slabs at the edge. Wipe away any slops of mortar on the liner before refilling the pond.

8

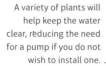

Stones offer access into and out of the water.

A variety of plants will help keep the water clear, reducing the need for a pump if you do not wish to install one.

9

Project 3

SMALL CONTAINER POND

For those who do not have the space or ability to dig a pond, a simple container, such as an old Belfast sink or similar vessel, makes the perfect alternative. You'd be surprised by how much wildlife a tiny pond attracts – from amphibians to dragonflies, and other insect larvae – and it also provides drinking opportunities for thirsty mammals.

YOU WILL NEED

Small, weatherproof container, such as a Belfast sink

Plug, small piece of pond liner, or sturdy plastic sheet

Bathroom sealant

Spirit level

Trowel

Bricks

Stones

Plants (see p.54)

Container pond gallery
An old sink (near left) looks charming and provides space for plants. An old bath (far left, top) sunk into the ground makes a spacious container pond. A wooden barrel (far left, bottom) offers a rustic effect.

To make your container pond as wildlife-friendly as possible, add a range of marginal and oxygenating plants, which will thrive in the relatively shallow conditions, and keep the water cleaner. If the vessel for the pond has steep sides, place bricks and other materials around the edge of the pond, both inside and out, to make an access ladder so amphibians and small mammals can enter and exit the water easily.

Position your container pond in partial shade – ideally so it gets sun in spring but some shade in summer. This will ensure the water warms up in the colder months but doesn't dry out when it's hot. You could also sink your container into the ground, which will help maintain a more even water temperature and allows amphibians, birds and mammals to use the pond easily – though you will still need to provide a way out.

SUGGESTED PLANTS

MARGINALS: **Water forget-me-not**
(*Myosotis scorpioides*)
FLOATING PLANTS: **Water soldiers**
(*Stratioites aloides*), **Frogbit**
(*Hydrocharis morsus-ranae*)

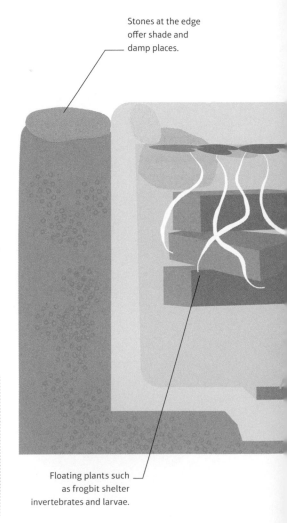

Stones at the edge offer shade and damp places.

Floating plants such as frogbit shelter invertebrates and larvae.

Sinking the pond into the ground keeps the water temperature more even, avoiding extreme fluctuations.

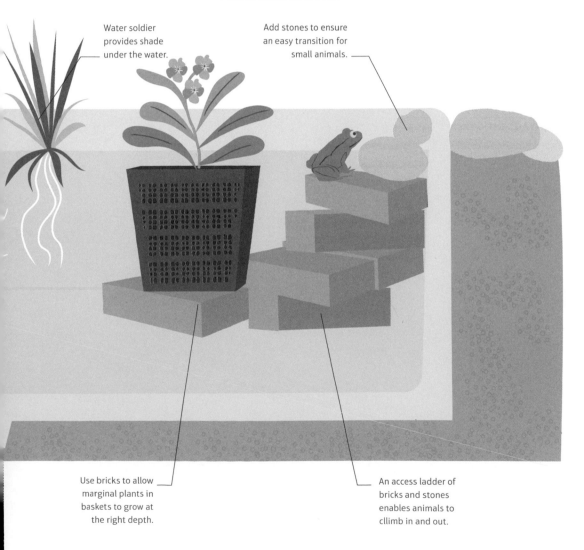

Water soldier provides shade under the water.

Add stones to ensure an easy transition for small animals.

Use bricks to allow marginal plants in baskets to grow at the right depth.

An access ladder of bricks and stones enables animals to cllimb in and out.

SMALL CONTAINER POND STEP-BY-STEP

STEP 1

If using an old sink, fit a plug in the plug hole and secure it in place with sealant. Otherwise, cut a piece of pond liner or plastic sheet to size so that it fits snugly in the bottom of your container, then apply sealant around the edges to make it watertight. Leave the sealant to dry following manufacturer's instructions.

STEP 2

Partially fill the pond with water to check that it's watertight. Even the slightest crack in the seal will let some water out. Leave for a few hours and check beneath the container for leaks. Drain, dry, and add more sealant if necessary.

STEP 3

To sink the container into the ground, trace the outline of the container into the soil with a trowel. Move the container to one side and dig a hole just shy of the total depth of the container, so it will sit slightly proud of the soil surface. This will enable wildlife to use the pond easily, while still visually working as a "container pond". You can also fully submerge the container if you want a more natural look. Check with a spirit level and adjust if needed.

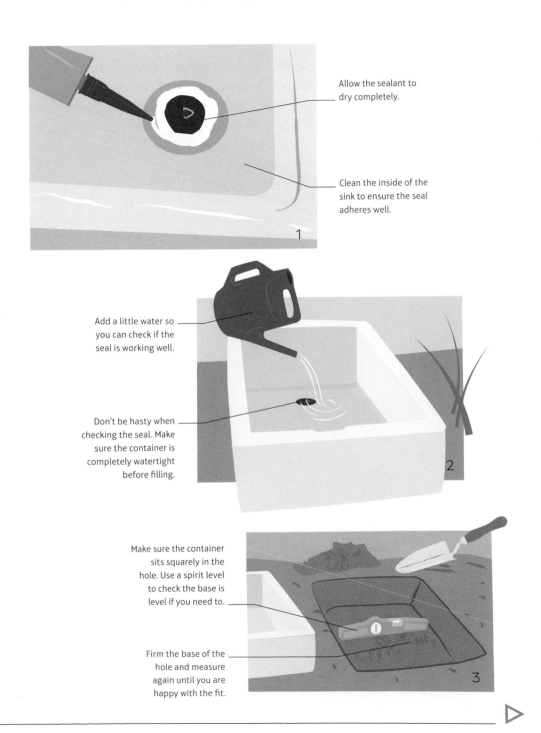

Allow the sealant to dry completely.

Clean the inside of the sink to ensure the seal adheres well.

1

Add a little water so you can check if the seal is working well.

Don't be hasty when checking the seal. Make sure the container is completely watertight before filling.

2

Make sure the container sits squarely in the hole. Use a spirit level to check the base is level if you need to.

Firm the base of the hole and measure again until you are happy with the fit.

3

STEP 4

Place the sink into the hole and check again that it is level and stable. Add bricks inside the sink to support your marginals. These will also create different water depths in the pond, making it a more appealing habitat for wildlife. Stand your marginals in their planting baskets on the bricks, checking they sit at the right depth – the top of the planting basket should sit just below the rim of the pond so it will be under the water.

STEP 5

Fill your pond, ideally with rain water, taking care not to disturb the marginal plants, then add floating or oxygenating plants. Overall you should aim for around two-thirds of the pond's surface to be covered with plant life. Frogbit and water forget-me-not are perfect in a container pond.

STEP 6

Place stones or bricks at one edge of the pond both inside and out to provide access into and out of the container. Stones will act as a ladder and as shelter for amphibians and invertebrates when they are not in the water. If the pond is not sunk into the ground, stack bricks or other materials against the outside and inside of the container for access. You may want to be creative and add an old, thick fishing rope, or chimney pots, rather than bricks, or use a log as a ramp.

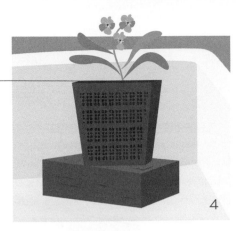

Check the planting baskets sit below the rim of the container before filling your pond.

Move plants around until you are happy with their placement.

Place floating plants such as water soldier and frogbit into the water.

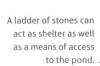

A ladder of stones can act as shelter as well as a means of access to the pond.

WILDLIFE RATING 5/10   TIME TO BUILD 1 day   EASE OF MAINTENANCE 7/10

Project 4
TROUGH POND

While frogs will spawn in virtually any shallow body of water, including puddles, toads are more fussy, and seem to prefer deeper ponds, often with fish present. Not all of us can make a large, deep toad pond in our gardens, but some of us may have room for a trough pond, which could provide the right habitat – as long as you can provide easy access up to it.

YOU WILL NEED

Metal livestock water trough or other large container, at least 60cm (2ft) tall

Bathroom sealant (optional)

Bricks

Planting baskets

Plants (see p.63)

Trough pond gallery Placing your trough beneath a window (near left) makes a feature of its planting. A trough pond (far left, top) offers a glimpse into the world below the surface. Surrounded by plants (far left, bottom), this trough blends with the garden.

Most container ponds are quite small, but you can create a much larger, deeper pond using an old water tank or trough, such as a livestock trough. These are long and deep, offering a completely different habitat to those you would find in a natural wildlife pond (see pp.32–41).

In this larger version of the container pond you would expect to see breeding newts and even toads, as these seem to prefer to breed in larger bodies of water. The increased depth will also enable you to make more of a feature of the planting. You can grow a wider variety of beautiful pond plants than is usually possible in a smaller container pond, including water lilies, which can form a striking focal point.

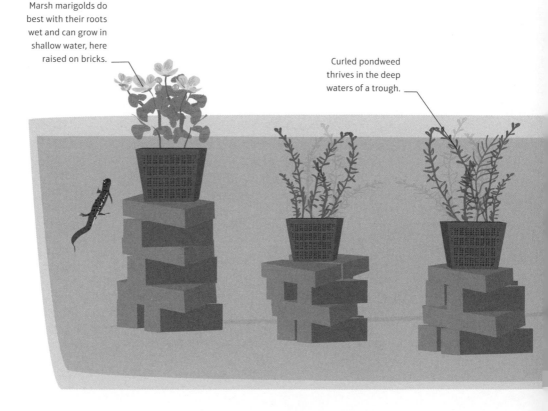

Marsh marigolds do best with their roots wet and can grow in shallow water, here raised on bricks.

Curled pondweed thrives in the deep waters of a trough.

Troughs offer a completely different habitat to a natural wildlife pond.

SUGGESTED PLANTS
———

BOG PLANTS: **Marsh marigold**
(*Caltha palustris*)
OXYGENATORS: **Curled pondweed**
(*Potamogeton crispus*), **Hornwort**
(*Ceratophyllum demersum*)
FLOATING PLANTS: **Water lily**
(*Nymphaea*)

Place water lilies on a pile of bricks, and take away a brick at a time as the llily grows.

Toads seem to prefer to breed in larger, deeper bodies of water.

A log pile doubles up as an access ladder and an additional habitat.

TROUGH POND STEP-BY-STEP

STEP 1

Position your trough in its final location – when it's full you won't be able to move it. It should sit in an area of partial shade, so that it ideally gets sun in spring but a bit of shade in summer. This will ensure the water warms up in the colder months, but doesn't get too hot later on.

STEP 2

Partially fill the pond with water to check it's watertight. If the water starts to leak, empty the water out and plug any holes or cracks with sealant, then leave to dry according to the manufacturer's instructions.

STEP 3

Add bricks to the bottom of the trough to support water lilies (see also Step 4). Stack more bricks at one side of the trough to create a shallow end of the pool, where birds will be able to bathe safely and amphibians will be able to enter and exit easily.

Look for a happy medium: a spot that gets some sun but won't be too exposed at the height of summer.

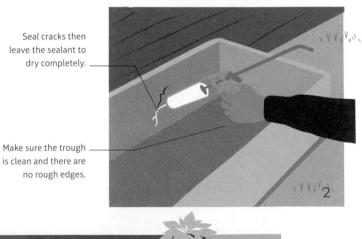

Seal cracks then leave the sealant to dry completely.

Make sure the trough is clean and there are no rough edges.

Bricks form steps at one end of the trough for access in and out.

Roughly position bricks to support floating plants in pots.

STEP 4

Place the water lilies, in planting baskets, on the piles of bricks. These plants need to sit just below the surface of the water initially, and can gradually be lowered by removing one brick level at a time, as the stems start to grow. Eventually, water lilies can sit in the bottom of the trough.

STEP 5

Plant up your marginal and bog plants in baskets, then stand them at the shallow end of the trough. Make sure they sit at the right depth (see pp.72–73). Fill your pond, ideally with rain water, ensuring the marginal plants sit securely when the trough is full. Finally, add the oxygenating plants, which float just beneath the surface.

STEP 6

Once the pond is filled and the plants are secure, create an access ladder by piling bricks or other materials, such as logs, around the outside of the container. This will allow the toads to climb up to the pond easily.

Ensure planting baskets are firmly on their bricks so they won't become dislodged.

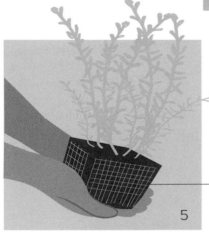

Lower plants in their baskets carefully into the water on to bricks at the correct planting depth.

Here, bricks provide a way out for animals, while logs form a ladder outside the trough.

Adjust the position of plants as you wish.

PLANTING UP YOUR POND

POND PLANTS

Plants are the most important feature of a pond. Each has the potential to support something different, adding to the enormous multi-layered ecosystem your pond will become. While we create shallows to enable birds to bathe, invertebrates to lay eggs and mammals to drink, we dig different depths largely so we can grow more plants to create the widest range of micro-habitats possible.

The best wildlife pond has a good mix of floating plants, marginals, oxygenators and bog plants (see pp.74–87), each of which thrives in a particular zone or depth of water (see pp.72–73). This mixture provides habitats for different species, from tadpoles and aquatic larvae sheltering from predators in rafts of marginals, to newts that wrap their eggs in leaves, to bees that land on lily pads to safely drink from the water. Every plant in your pond is an opportunity for some species or other – the more you plant, the more wildlife arrives.

Some of my plants are doing better than others; that's perfectly normal. As your pond matures and finds its own way, the plants that grow in it will do so, too.

The ones that thrive in my allotment pond differ from those that do best in my garden pond. Planting the widest range of plants possible and seeing which ones "take" is the best option here. My favourites tend to be the floaters. I like water lilies (although sadly my ponds are too small for them), water soldiers and frogbit. Marginals like water forget-me-not and brooklime are just as beautiful. Growing pond plants with tall, emergent stems enables dragonfly and damselfly nymphs to climb out of the pond for their final metamorphosis.

Floating plants
A mix of different water lilies forms an elegant focus, and provides shade under the water too.

Pond planting zones indicate how deep a plant needs to be planted. Pond plant nurseries tend to split the zones into five, as shown here, to help you when planting up your pond. In essence there are only three true zones: bog plant/wetland; shallow water; and deep water. Bear in mind that most planting zones overlap: floating and deep-water plants may also be oxygenators, while many marginals also do well as bog plants. The way you plant your pond depends on the type of

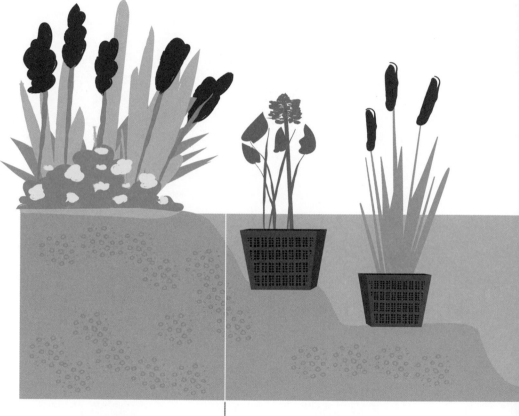

ZONE 1
BOG PLANTS

ZONES 2 AND 3
MARGINAL PLANTS

pond you have. In a wildlife pond, you can plant into the subsoil, which sits on top of the liner (see pp.32–41). In a more formal pond with shelves, or in a container or trough, it's best to use planting baskets. The overall effect is similar – once the plants mature you shouldn't be able to see any pond baskets. Some pond plants can be dropped into the water, where they'll float or root into the substrate.

I tend to grow only native pond plants (see pp.176–177), as they are unlikely to cause problems if they escape into the wild, unlike some non-natives that may become invasive (see pp.88–89). Some natives such as yellow-flag iris, water starwort and water soldiers are also vigorous and are not ideal for small ponds.

POND PLANTS AND SOIL NUTRITION

Most pond plants are vigorous and do not need soil that is highly nutritious. Garden topsoil tends to be rich in nutrients, which not only makes it unsuitable for pond plants, but may also encourage too much algae to bloom on the water's surface, smothering other plants and aquatic life in the pond. It's far better to use low-nutrient aquatic compost or subsoil in your pond and planting baskets.

ZONE 4
OXYGENATORS AND
DEEP WATER PLANTS

ZONE 5
FLOATING PLANTS

BOG PLANTS

Bog plants grow happily around the pond edge, taking advantage of the damp growing conditions found here. Many are native to wet meadows, while some will cope with slightly drier conditions when the water level drops in summer.

Bog plants can add an extra layer of interest and beauty to your pond – many are absolutely gorgeous – while also providing important cover from predators for animals entering or leaving the water.

Around my garden pond I grow cuckoo flower, primrose and cowslips, ragged robin, water figwort and purple loosestrife. These flower at different times of year, providing pollinators with a long season of nectar and pollen. Some, such as the primrose and cuckoo flower, are used as food plants by butterflies and moths. Others have different specialities – water figwort is particularly loved by wasps (of which there are many wonderful species besides the common wasp, which has such a bad reputation). Purple loosestrife is the absolute bee's knees when it comes to providing food for pollinators such as bumblebees, honey bees and hoverflies. These pond-edge plants make my pond and garden more interesting, more vibrant and more wildlife-friendly. I wouldn't be without them.

> Bog plants enhance your pond habitat, catering for pollinators and other invertebrates.

MARSH MARIGOLD
Caltha palustris

HEIGHT 40cm (16in) **SPREAD** 50cm (20in)
FULL SUN

Marsh marigold has glossy, heart-shaped leaves and bright yellow spring flowers – what's not to like? Bees love the flowers, too.

CUCKOO FLOWER
Cardamine pratensis

HEIGHT 45cm (18in) **SPREAD** 30cm (12in)
PARTIAL SHADE

Pretty, lilac flowers appear from late spring, above attractive, divided foliage. This is the food plant of the orange-tip butterfly; its caterpillars nibble on the leaves and flowers before pupating.

HEMP AGRIMONY
Eupatorium cannabinum

HEIGHT 1.5m (5ft) **SPREAD** 1.5m (5ft)
FULL SUN TO PARTIAL SHADE

One of the best plants you can grow for butterflies, the flowers of hemp agrimony are packed with nectar. Its hemp-like leaves are attractive, too, although they may inspire a few questions from guests!

MEADOWSWEET
Filipendula ulmaria

HEIGHT 90cm (36in) **SPREAD** 30cm (12in)
FULL SUN TO PARTIAL SHADE
Meadowsweet bears gorgeous, fluffy,
cloud-like flowerheads in summer, which are
loved by pollinators. It also has attractive, fresh
green foliage.

RAGGED ROBIN
Lychnis flos-cuculi

HEIGHT 75cm (30in)
SPREAD 75cm (30in)
FULL SUN TO PARTIAL SHADE
The prettiest ragged, pink, star-shaped flowers
appear from late spring. Dot them among
grasses for a natural, wild look.

PURPLE LOOSESTRIFE
Lythrum salicaria

HEIGHT 2m (6½ft)
SPREAD 60cm (24in)
FULL SUN TO PARTIAL SHADE
Purple loosestrife bears tall, purple
flower spikes in summer. Bees and
other pollinators love it.

COWSLIP
Primula veris

HEIGHT 25cm (10in)
SPREAD 25cm (10in)
PARTIAL SHADE

A versatile primula, it can cope with drier soils but looks lovely around a pond in spring. Clusters of funnel-shaped, yellow blooms appear on tall stems above crinkly leaves.

WATER FIGWORT
Scrophularia auriculata

HEIGHT 80cm (32in) **SPREAD** 50cm (20in)
FULL SUN TO PARTIAL SHADE

This beautiful plant has square stems from which tiny maroon flowers appear from June to September. It's fantastic for pollinators, providing nectar for both bees and wasps.

COMMON VALERIAN
Valeriana officinalis

HEIGHT 1.5m (5ft) **SPREAD** 1m (3ft)
FULL SUN TO PARTIAL SHADE

The nectar-rich summer flowers of common valerian attract a wide range of pollinators, and the foliage is pretty, too. A soothing tea can be made from its roots.

MARGINAL PLANTS

Marginal plants grow around the edges (or margins) of your pond, where the water is shallow. They "like having their feet wet", which typically means having their roots in water and the majority of the plant above the shore line.

Marginals help to soften the pond edge and also clean the water of impurities. They contribute to making a "wetland" habitat around your pond, which in turn attracts a wide range of wildlife, including butterflies and moths, which feed on and pollinate marginals.

In naturally occuring ponds, marginals self-seed into the mud at the pond edge.

If you dig a wildlife pond (see pp.32–41) you can mimic this effect by planting directly into the muddy edges you've created. Alternatively, plant your marginals in pond baskets and place these on your shallow shelves, where the water is no more than 15cm (6in) deep.

Marginal plants offer some of the best habitats for wildlife, and some will "raft" across the surface of the pond, providing shelter and shade for a huge range of species. Tadpoles and other larvae will hide from predators among floating stems and leaves, while newts lay their eggs on them, wrapping each egg within an individual leaf. You might even find frogs basking here on hot sunny days.

> Margins warm up quickly in spring, perfect for amphibians and invertebrates.

WATER PLANTAIN
Alisma plantago-aquatica

HEIGHT 1m (3ft)   **SPREAD** 50cm (20in)
SUN OR PARTIAL SHADE
With a tendency to self seed, water plantain is
best for large ponds. Tall sprays of pretty white
flowers grow in summer from rosettes of large,
oblong leaves.

FLOWERING RUSH
Butomus umbellatus

HEIGHT 1.5m (5ft)   **SPREAD** 50cm (20in)
FULL SUN
Flowering rush bears clusters of pastel pink
flowers that attract pollinating insects in summer.
Give it plenty of room to grow, and divide
regularly to encourage flowering.

SQUARE-STEMMED ST JOHN'S WORT
hypericum tetrapterum

HEIGHT 30cm (12in)   **SPREAD** 30cm (12in)
FULL SUN OR PARTIAL SHADE
This unusual perennial with square stems and
yellow flowers in summer is happy in the pond
margins or a bog garden. A citrusy scent is
released when you rub the leaves.

YELLOW-FLAG IRIS
Iris pseudacorus

HEIGHT 1.2m (4ft)   **SPREAD** 1m (3ft)
FULL SUN
Best for large ponds, this iris bears sword-shaped
leaves and yellow flowers in late spring. Dragon-
and damselflies use the emergent stems to climb
out of the water for their final metamorphosis.

WATER MINT
Mentha aquatica

HEIGHT 30cm (12in)
SPREAD 50cm (20in)
FULL SUN OR PARTIAL SHADE
This marginal provides
egg-laying opportunities for
newts. Its scented leaves give
rise to globe-like, lilac-pink
flowerheads, which are loved
by bees.

BOGBEAN
Menyanthes trifoliata

HEIGHT 50cm (20in)
SPREAD 1.5m (5ft)
FULL SUN
Another one for a large pond, bogbean can
spread very quickly. It bears emergent, oval leaves
and fluffy, star-shaped white flowers in summer.

WATER FORGET-ME-NOT
Myosotis scorpioides

HEIGHT 30cm (12in)   **SPREAD** 50cm (20in)
SUN OR SHADE
This marginal provides shelter for tadpoles, while
newts lay their eggs in the leaves. Blue flowers
appear from May to July, providing nectar and
pollen for pollinators.

AMPHIBIOUS BISTORT
Persicaria amphibia

HEIGHT 70cm (28in) **SPREAD** 50cm (20in)
SUN OR SHADE
Happy at the pond's edge or submerged, this
perennial has large, oblong leaves that shelter
invertebrate larvae and create shade, helping to
reduce algae. Its pink flowers attract pollinators.

ARROWHEAD
Sagittaria sagittifolia

HEIGHT 1m (3ft) **SPREAD** 1.5m (5ft)
SUN OR PARTIAL SHADE
With emergent leaves shaped like arrowheads
and oval leaves that float on the water's surface,
this marginal makes a beautiful addition to
ponds. Small white flowers appear in summer.

BROOKLIME
Veronica beccabunga

HEIGHT 10cm (4in)
SPREAD 1.2m (4ft)
FULL SUN OR PARTIAL SHADE
This beautiful marginal rafts
across the surface of the pond,
creating shade and shelter
wherever it grows. Its leaves are
also used by egg-laying newts.
Small, pretty flowers grow on tall
stems from late spring to autumn.

OXYGENATORS AND DEEP-WATER PLANTS

Oxygenators are technically deep-water plants. You usually don't need to plant them; they root into the mud and detritus at the bottom of the pond. While all plants oxygenate water to some degree, so-called oxygenators are considered to be especially efficient at it.

Oxygenators provide a unique habitat for invertebrates far below the surface, and some of them knit together over time to create what I think of as a kelp-like underwater forest. Here invertebrates hang out at different levels, some of them laying eggs, some of them chasing prey while others are hiding from predators.

There are several types of oxygenator to choose from. Some can be fussy, while others may be overly vigorous, so it pays to add them to the water one at a time to see which fares best. In my pond, curled pondweed and spiked water milfoil reign, while hornwort, which I prefer, barely gets a look-in.

Many oxygenators are sold in bunches, rather than in pots. You simply throw them into the water and leave them to establish by themselves where they are happiest.

Some oxygenators form a kelp-like underwater forest.

COMMON WATER STARWORT
Callitriche stagnalis

PLANTING DEPTH 30–100cm (12–39in)
SPREAD 1m (3ft)   **FULL SUN OR PARTIAL SHADE**
This plant roots in mud at the bottom of the
pond. Small, white star-shaped flowers appear
from May to August. Throw clumps into the water
or plant them in a basket.

HORNWORT
Ceratophyllum demersum

PLANTING DEPTH 30–100cm (12–39in)
SPREAD 1.5m (5ft)   **SUN TO SHADE**
One of the most common oxygenators, hornwort
has free-floating, branched stems of dark green
foliage. Throw bunches of hornwort into the
water where it will establish.

WILLOW MOSS
Fontinalis antipyretica

PLANTING DEPTH 30–100cm
(12–39in)   **SPREAD** 30cm (12in)
SUN OR PARTIAL SHADE
This oxygenator is evergreen with
branched stems. It grows beneath
the surface, in mats, and prefers
moving water.

MARE'S TAIL
Hippuris vulgaris

HEIGHT 20cm (8in)
SPREAD 50cm (20in)
FULL SUN

This attractive oxygenator bears tall whorls of mid- to dark green leaves above the water surface. Plant into the mud or a basket.

MARSH PENNYWORT
Hydrocotyle vulgaris

HEIGHT 10cm (4in)   **SPREAD** 2m (6½ft)
SUN TO PARTIAL SHADE

This small oxygenator has pink flowers in summer. It grows in boggy areas and margins as well as in water up to 10cm (4in) deep. Plant into the mud or a basket of aquatic compost.

SPIKED WATER MILFOIL
Myriophyllum spicatum

PLANTING DEPTH 30–100cm (12–39in)
SPREAD 1.5m (5ft)   **FULL SUN**

Spiked water milfoil forms a dense mat of beautiful, feathery foliage. Plant this oxygenator into the mud or a basket of aquatic compost.

CURLED PONDWEED
Potamogeton crispus
PLANTING DEPTH 30–100cm (12–39in)
SPREAD 1.5m (5ft) **FULL SUN TO PARTIAL SHADE**
Brown-green wavy-edged leaves remain beneath
the surface of the water, rooting into mud and
knitting together, like kelp. Drop into the water or
plant in a basket of aquatic compost.

FENNEL PONDWEED
potamogeton pectinatus
HEIGHT 10cm (4in) **SPREAD** 1m (3ft)
FULL SUN TO PARTIAL SHADE
Submerged, grass-like leaves gradually form a
dense mass that resembles a meadow just
beneath the surface. Plant fennel pondweed in
baskets on shelves 50cm–2m (20in–6ft) deep.

COMMON WATER CROWFOOT
Ranunculus aquatilis
PLANTING DEPTH 5–30cm (2–12in)
SPREAD 50cm (20in) **FULL SUN**
Feathery foliage forms mats
beneath the surface, while pretty,
daisy-like flowers grow above, in
summer. Plant into a basket of
aquatic compost or weigh down in
the mud with a stone or weight.

FLOATING PLANTS

When we think of floating plants, most of us imagine water lilies. Technically deep-water floating-leaf plants, water lilies grow in planting baskets at the bottom or on lower shelves on the pond, their long stems and plate-like leaves floating on the water's surface.

Not all ponds are suitable for water lilies as they're typically found in larger bodies of water, although you can buy dwarf species for small ponds. As well as our native water lily, there are also plenty of cultivars and tropical species which may be better suited to garden ponds. You may spot baby frogs sitting on lily pads in summer, or honey bees, which land on them to drink from the pond.

Free-floating plants float without rooting into mud. These include water soldiers, which resemble the head of a pineapple and sit erect just below the water's surface, like soldiers standing to attention. Frogbit is a lovely free-floating plant that looks like a tiny water lily.

WATER VIOLET
Hottonia palustris

PLANTING DEPTH 10–60cm (4–24in)
HEIGHT 50cm (20in) **SPREAD** 1m (3ft)
PARTIAL SHADE

Star-shaped pink-white flowers emerge from fern-like foliage from late spring. Plant in the shallows initially, and it will spread its green mats of foliage deeper into the pond if it's happy.

FROGBIT
Hydrocharis morsus-ranae

PLANTING DEPTH 5–20cm (2–8in)
HEIGHT 10cm (4in) **SPREAD** 1m (3ft)
FULL SUN TO SHADE
Small, water-lily-like leaves float on the surface,
while pretty white and yellow flowers rise above
them in mid- to late summer.

WATER LILIES
Nymphaea spp.

HEIGHT 75cm (30in) **SPREAD** up to 1.5m (5ft)
FULL SUN
Native species flower in white in summer, above
plate-like leaves. Check planting depths: most
should be 30–75cm (12–30in) below the water's
surface, and do best in very large ponds.

BROAD-LEAVED PONDWEED
Potamogeton natans

PLANTING DEPTH 30–100cm (12–39in)
HEIGHT 10cm (4in) **SPREAD** up to 2m (6½ft)
FULL SUN OR PARTIAL SHADE
Pretty, oval leaves sit on the surface while
insignificant, knobbly flower spikes appear in
summer. This plant can be difficult to establish,
but may become rather rampant over time.

WATER SOLDIER
Stratiotes aloides

PLANTING DEPTH 20–100cm (10–39in)
HEIGHT 10cm (4in) **SPREAD** 1m (3ft)
FULL SUN OR PARTIAL SHADE
Rosettes of sharp, sword-shaped leaves sit just
below the surface. Pretty white flowers appear
from the centre.

INVASIVE PLANTS

Invasive plants are native to other parts of
the world, where their growth is checked
by their local environments. In non-native
areas they can out-compete other plants.
They can be particularly troublesome if
they escape from ponds, causing problems
in rivers, streams and other wetlands.

I grow only native pond plants as they
are attractive to native wildlife and are
unlikely to disrupt natural ecosystems if
they escape into the wild (see pp.176–
177). If you want to add non-natives to
your pond, bear in mind that some are
very vigorous; check this before planting.

The five plants shown here are
considered the worst invasive pond
plants in the UK and were banned from
sale (see also p.183). Always buy pond
plants from a reputable supplier. It's a
good idea to "quarantine" new plants
before putting them in your pond. Place
them in a bucket of water for a few days
and then check them for new growth of
plants other than the ones you bought.

FAIRY MOSS
Azolla filiculoides

Native to North America, this water fern forms
mats of foliage so dense that it blocks light and
can starve the water of oxygen, threatening all
life in the water.

AUSTRALIAN SWAMP STONECROP (NEW ZEALAND PIGMYWEED)
Crassula helmsii

Another mat-forming pond plant, this Australia and New Zealand native quickly shuts out light to ponds, which can kill virtually all life within. The plant bears fleshy leaves and small white flowers.

FLOATING PENNYWORT
Hydrocotyle ranunculoides

This American native has fleshy stems and round, toothed leaves. Watch out – it's sometimes mistakenly sold as "marsh pennywort", a European native. It can starve waterways of oxygen, clog drainage systems, out-compete native plants, and reduce biodiversity.

WATER PRIMROSE
Ludwigia grandiflora

Despite its pretty yellow flowers, the dense foliage of water primrose can clog waterways and contribute to flooding. It is native to the Americas.

PARROT'S FEATHER
Myriophyllum aquaticum

Once a popular pond plant, this South American native may still be sold (illegally) as Brazilian water milfoil. It forms such dense mats of foliage that it can alter the flow of rivers.

OBSERVING YOUR POND

THE POND THROUGH THE YEAR

I like to think of ponds as having a life of their own. You dig one, fill it with water and plant it up, and then set it free. It might turn green initially, before gradually acclimatizing to its surroundings. Plants will grow (and some will die), insects and other invertebrates will arrive.

As the pond matures it becomes part of the garden, home to a variety of different animals, each of which brings a new dimension, a new complexity, to that little body of water outside your back door. It changes with the seasons, and is at its busiest in spring, of course, with a riot of amphibians and invertebrates descending to mate and lay eggs in the water. In summer you'll find the larvae of frogs, toads and newts, plus a million things you can't identify – wriggling things on the surface, beetles zipping in and around stems and roots in the shallows, and intriguing nymphs of bigger insects.

Autumn is quiet and winter is silent. You'll still see birds bathing, the odd mammal popping in for a drink, but otherwise nothing. Invertebrates and amphibians shelter in quiet corners, beneath log piles, in the mud, or tucked up amid tussocky grass. In the pond, too, things are quietly sitting out the colder months below the surface before rising to party again, when spring returns.

Height of summer
In summer your pond is a busy place, with larvae developing and pond plants buzzing with pollinators.

EXPLORING POND LIFE

Taking time to sit and look at your pond is extremely rewarding at any time of year. If you regularly watch your pond, you'll see how it changes through the seasons. You can start to explore interactions both above and beneath the surface.

Ponds are fascinating, wondrous things, and getting to know all the wildlife that uses them is the very best thing about having one. You might hear the plop of a frog dropping into the water or spot dragonflies mating, before the female lowers her abdomen to lay eggs among pond plants. You could see pond snails eating algae just beneath the water, and backswimmers and diving beetles rising up for air while pond skaters dance across the surface. Sit in the evening with a glass of wine and bats might flit above you, catching mosquitoes, or the rustle of foliage might alert you to a hedgehog popping in for a drink.

Sitting by the pond and enjoying it from the land might be enough for

View from above
Look regularly into your pond and you will notice all the small (and larger) changes over time.

you, but discovering it in more detail is fascinating, especially if you explore below the surface. Why not shine a torch into the water on a summer's evening? You may spot newts, which are more active at night than in the day. You might want to invest in a trail camera or two, positioning them around the pond to catch birds bathing and insects laying eggs, or even a waterproof camera that you can lower into the pond to see what's in the water – try this in spring when amphibians are breeding or in summer when the water is thick with tadpoles.

Pond dipping (see pp.96–99) is a great activity to get to know your pond, too, and perfect for kids to join in.

Underwater world
An underwater camera can show you what's happening in your pond.

You'll find all sorts of creatures (see pp.100–102), including diving beetles, backswimmers, dragonfly, damselfly and mayfly larvae, as well as caddisfly cases. All have a role in the pond, from keeping the water clean to helping numbers of other species. This complex web of predator and prey that exists in your pond contributes to its role as a central life source in your garden. Arm yourself with a microscope and you'll enter a whole new world of tiny organisms invisible to the naked eye (see p.103).

POND DIPPING

To observe the ecosystem of your pond up close, try pond dipping. You can pond dip at any time of year, although spring and summer will yield more results as the water is rich with new invertebrate life. Pond dipping is great for all ages, but it's a particularly good activity to do with kids.

When pond dipping with children, ensure they know the essential rules (see opposite), and that they are always with an adult. You don't need a lot of kit. Use a fishing net or even an old kitchen sieve, and white trays to help you see what you find. You may also want to record species – taking note of the day, time and year can help you work out how quickly new species are colonizing your little body of water.

The web of life in a pond is complex, and includes microscopic species, detritivores that eat dead plant material, algae-eaters, predators and parasites. I've listed a few species you might find (see pp.101–103), but you may wish to invest in a freshwater life book to cover species at all stages of their lifecycles. Learning to identify these species will greatly enhance your enjoyment of your pond.

ESSENTIAL RULES
OF POND DIPPING

Always kneel to avoid falling in.
Never stand at the edge.

Use separate containers for predators
and prey – if you keep them together
you may find the prey keeps
"disappearing".

Avoid leaving the trays in the hot sun,
which can cause the water to overheat
and harm your catch.

Return all of your catch to the
pond after you've finished.

Never move animals from one
pond to another.

Rest the net in the pond while you
investigate, to avoid harming anything
that is caught in it.

Looking at what you find
Take your time to study
what you find in your
container, moving plants
and animals around
gently if you need to
get a better view.

Approach the pond slowly and carefully – the tiniest of vibrations from your feet landing on the earth will be detected by animals in the pond so it's worth sitting quietly at the pond edge for a few minutes after you've arrived. Find a sturdy, non-slippery spot to kneel safely at the edge and scoop some water into your container as a starting point. Make a sweep of the pond with your net, submerging it fully and then emptying it into a container. Avoid the temptation to fill your tray with vegetation as too much will create hiding places for small animals, which you then may not be able to find. While looking in the tray, rest your net in the water so anything trapped in the net can escape easily.

At each stage, work slowly and carefully so as not to harm the animals.

Try different areas and depths of the pond to catch the greatest range of species. Be kind – don't keep anything in the trays for longer than is necessary, and be particularly careful on a hot day when the water in your tray will heat up quickly. Separating predators from prey (using a second tray) is also a good idea: keep dragonfly and damselfly nymphs, diving beetle larvae and backswimmer nymphs in a separate tray or they will make short work of tadpoles and other larvae. Once you've finished observing and recording, gently return your catch to the water.

> Work slowly and carefully so as not to harm the animals.

YOU WILL NEED

Long-handled fishing net with a 1–2mm (1/16in) mesh

At least two white, wide-bottomed containers, such as old baking trays or ice-cream tubs

Magnifying glass (optional)

ID guide (see also p.183)

Notepad and pencil

Dip your container in the water, partially filling it with pond water. Set it down on the side of the pond.

Gently sweep the net beneath the surface of the water. Lift it out and empty the contents into your container.

Remove most of the vegetation as this may obscure the animals.

As the water settles, start to look at the animals wriggling in the water. Identify and record what you have found.

LESSER-KNOWN POND LIFE

There's a huge array of life in your pond.
Each species has its own role; there are no
"bad" or undesirable ones, except perhaps
for fish, which eat so many other life forms.
Embrace everything that arrives naturally
– why not learn to identify it, too.

The more you know about your pond life, the richer your experience of your pond. As well as the tadpoles and pond skaters you might expect, there are other species that take more effort to find. These include bizarre, ancient, microscopic life forms such as rotifers and water bears, which withstand drought and live in dried-up ponds, dormant until the water returns, sometimes for years. They may arrive at your pond in this dehydrated state, having been blown in by wind.

Other pond life arrives on the leaves of plants or feet of birds – leeches, water hog lice, snails and worms may turn up this way. Some species you may spot in your pond-dipping tray will have been laid as eggs in the pond. They may live there for a few months or even years as larvae or nymphs, before turning into adults and flying off to lay their own eggs.

WHAT'S THE DIFFERENCE BETWEEN A LARVA AND A NYMPH?

The larvae and nymphs of a species are the immature stages in that invertebrate's lifecycle. As a general rule, larvae don't resemble the adult stage at all, and go through "complete" metamorphosis before becoming an adult. Nymphs look more like the adults and go through "incomplete" metamorphosis.

WATER HOGLICE
Asellus aquaticus

LENGTH 10mm (½in)
Looking like tiny, slightly flattened woodlice, these aquatic crustaceans are common in ponds. They feed on decaying leaves and silt and can survive in low-oxygen environments.

BLOOD WORMS
Chironomidae

LENGTH 25mm (1in)
These look like little red worms. They're the larvae of a family of non-biting midges known as chironomid flies and are an important food source for a number of other pond species. They feed on rotting organic matter.

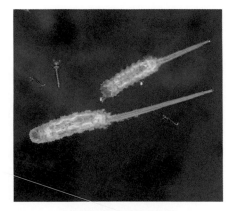

RAT-TAILED MAGGOTS
Eristalis

LENGTH 6.5cm (2½in), including tail
Easily the best-named of all the pond creatures, rat-tailed maggots are the larvae of hoverflies, most of which are in the Eristalis genus. They have adapted to aquatic conditions by breathing through a long tube, earning them their name.

LEECHES
Hirudinea

LENGTH 40mm (1½in)
There are lots of types of leech in ponds, such as *Erpobdella* (pictured). Don't be alarmed – only one species is capable of cutting through human skin and it's nearly extinct. Most eat other organisms such as fish, worms and tadpoles.

WATER MITES
Hydrachnidia

LENGTH 2mm (⅟₁₆in)
Found on backswimmers and beetles, on which they feed parasitically, water mites, such as *Eylais* (pictured), look like tiny spiders. They vary in colour, with some species in eye-catching shades of red, blue or yellow.

CADDISFLY LARVAE
Trichoptera

LENGTH 3–40mm (⅛–1½in)
Caddisfly larvae cover themselves with items from the pond for protection and camouflage. Different species, such as *Limnephilus* (pictured), use different items, such as plant roots, discarded snail shells, bits of grass or grains of sand.

DRAGONFLY, DAMSELFLY AND MAYFLY NYMPHS

LENGTH 5–60mm (¼–2½in)
Nymphs of dragonflies (above, and pp.134–135), damselflies (see pp.118–119) and mayflies are similar. Dragonfly nymphs are chunky with big eyes. Damselfly and mayfly nymphs are slender with feathery gills at the end of their body.

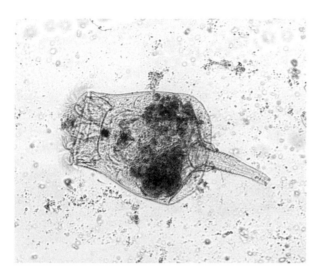

BDELLOID ROTIFERS
Bdelloidea

LENGTH up to 1mm (less than 1/16in)
Microscopic and invisible to the
naked eye, bdelloid rotifers draw
food particles into their body using
tiny, fast-moving hairs
called cilia.

HYDRA
Hydra

LENGTH 1mm (less than 1/16in)
These freshwater organisms consist of a long
body with tentacles that grab at passing food
particles. Some grow large enough to be seen
with the naked eye.

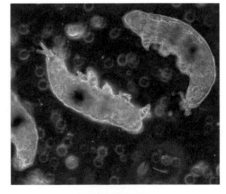

TARDIGRADES
Tardigrada

LENGTH up to 1.5mm (1/16in)
Also called water bears or moss piglets, owing to
their lumbering, mammal-like appearance under
a microscope, most tardigrades are found in
damp habitats like moss, but some live in ponds.

SPRING

Spring is the most magical time. Suddenly, after months of cold, rain, frost and ice, the garden comes alive. And it all starts with little stirrings in my pond.

The frogs are always the first to get the party started. They overwinter (hibernate) from around October to February or March, beneath compost heaps, in log piles or buried into the mud at the bottom of ponds. They emerge when temperatures hit around 5°C (41°F) – typically after rain – and head straight to their breeding grounds. In the British Isles frogs start spawning in the south west first, with spawnings moving across the country in a north-easterly direction. In a mild winter Cornish frogs may spawn as early as Christmas Day, but they usually start a couple of weeks later. After Cornwall and Devon, frogs spawn in south Wales and Hampshire, before spawning in Brighton, where I live, in the second week of February, ahead of the Midlands, the north and finally Scotland as late as April. I follow the action on Twitter, everyone excitedly exclaiming "the frogs are back!" It's a joyous time of year.

From the middle of February I creep into the garden each morning to see if I can spot signs of frogs in the water or on their way to it. I also check my allotment pond and the ponds of parks around the city – which

Frogspawn
One of the first signs of spring, frogspawn appears in ponds usually from February onwards.

Newt in the water
Male smooth newts develop a crest along their back in breeding season.

will be the first that frogs spawn in? When will I see the first blob? My usual running route, which takes me along Brighton seafront, is abandoned for the parks, allotments and community gardens of Brighton and Hove. It covers about six miles and takes in nine ponds. Nine ponds that gradually, over that beautiful transition from winter to spring, fill up with masses of horny amphibians.

Some years we have a dry spring, and the first rain of the season brings all the frogs out at the same time. This is my favourite type of spring. Ponds boil with ripples and croaks, as hundreds – sometimes thousands – of frogs arrive at once for a giant mating party that lasts just a few days. I watch "mating balls" of up to 10 frogs all rolling around together. In wet springs activity is more subdued, with frogs arriving gradually, over the course of a few weeks. Either way, ponds across the city eventually fill up with masses of frogspawn and I wait, again, for the first tadpoles to hatch.

We still get frosts in February, sometimes hard ones, and the frogspawn is frozen in the water. You might get a few blobs of spawn before the frost and the bulk of the spawning done afterwards; other times I've seen entire ponds packed with frozen frogspawn. Most of it survives; it's hardy stuff. Toads start spawning a couple of weeks after frogs. I watch them, too, and listen to their squeaks and water ripples. By the end of March ponds are full of wriggling masses of tadpoles, with newts arriving in the water to fatten up on them before

Ponds boil with ripples and croaks, as hundreds of frogs arrive.

their own breeding season begins. People often tell me there are too many tadpoles in their pond and they want to move them, or they're scared there will be "too many" frogs when it's time for them to emerge in summer. This never happens: far too many other species eat frog tadpoles, from newts and backswimmers to birds and dragonfly larvae. Some eat toad tadpoles too, despite them being slightly toxic. Newts are the last amphibians to breed, from late spring to early summer. Newt courtship takes place on the pond bottom, far below the prying eyes of excited wildlife gardeners – I've only caught sight of it twice.

Elsewhere in the pond, invertebrates like pond skaters and backswimmers are emerging and breeding, while dragonfly and damselfly nymphs crawl up emergent plant stems to finally metamorphose into adults. By the end of spring, ponds are lush and full, a world away from the cold, icy pools they were at the start of the season. I love this transformation. I basically live for it.

Pond skater
After emerging from hibernation, pond skaters lay their eggs.

Early spring is a riot. Frogs and toads descend on ponds for often huge mating parties, with their calls ranging from low rumbling croaks to loud squeaks and bird-like whistles, depending on the species and the number of individuals present. If you venture into the garden at night you'll hear the pond before you see it. Turn on your torch and a thousand eyes will greet you, lighting up little amphibian blobs splashing their way to the next generation.

Insects are busy, too, while pond plants are starting to grow. You might spot midges gathering above the water's surface in late afternoon, or the first leaves of marginal plants such as

Dragonfly nymph
This voracious underwater predator moults up to 14 times until it is fully grown.

brooklime and water forget-me-not creeping back to life. Birds continue to drink and bathe in the pond by day, while at night hedgehogs and frogs may drink from the pond as bats forage above it.

As soon as frogspawn is laid, the predators come out in force: the larvae of aquatic invertebrates such as dragonflies and damselflies hunt tadpoles in the water, as do adult and juvenile backswimmers and water beetles, fish and even grass snakes. Newts eat both tadpoles and the eggs in spawn, leaving empty jelly in their wake.

You might not be happy about your precious frogspawn being eaten, but it provides many species with a nutritious meal early in the year, usually after months of winter torpor spent without a food source to sustain them. Tadpoles are key to the next generation of frogs, yes, but they're also there to provide food for other species, as painful as it is to admit it.

Spring growth
Leaves of marginals start into growth around the pond edge in early spring.

If you venture into the garden at night in early spring, you'll hear the pond before you see it.

COMMON TOAD
Bufo bufo

The common toad is absolutely beautiful, with warty, olive-brown skin and amber eyes. Unlike the common frog (see pp.166–167), it crawls rather than hops, and its dry skin means that it doesn't need to spend as much time in water – you're more likely to find one in a log pile than in your pond.

The fourth most common amphibian in Europe, the common toad is absent only from Iceland, Ireland, northern Scandinavia, and a few Mediterranean islands. It's found in a variety of habitats, including gardens. Adults eat a number of invertebrates, such as ants, beetles and small slugs and snails.

BREEDING

This takes place in spring in quite large, deep bodies of water, often with fish present. Common toads typically breed in the pond where they started life. Mating

Warty, dry skin enables toads to spend time far away from water outside the breeding season.

Look out for amber eyes with horizontal slits.

Common toad
Poison glands give toads an advantage over other amphibians, as predators tend to avoid eating them.

MALE AND FEMALE TOADS
Toads head to their ancestral breeding ponds to mate, often following the same route every year. The male is much smaller than the female and may climb on to her back as she heads to the pond.

takes place in the water with the male clasping the back of the female. This can last for a few hours before the female releases her long ribbons of eggs into the water, and the male fertilizes them as they emerge. Sometimes huge mating balls can form, with several males vying for (and attaching themselves to) one female, who can occasionally drown.

Toad tadpoles are black and often school in groups on the surface of the water, like fish. They eat algae in the pond before starting to feed on tiny invertebrates such as water fleas. Once their legs have developed, in summer, they emerge from the pond.

TOAD TOXINS

Like most toads, the common toad is slightly poisonous, with glands in its skin containing toxins. Many predators therefore avoid eating the adults or their offspring. If you (or your curious dog) come into contact with toad toxins, it may cause slight foaming at the mouth and a mild tummy ache. This is nothing to worry about but is a good reason to avoid picking up toads. If you have to pick them up – for example to save one from being run over – always use gloves.

TOAD SPAWN
Laid in ribbons, toad spawn is fertilized by the male as the female lays it. She wraps the ribbons around submerged plant stems. It hatches into tadpoles after about a week, depending on the weather.

TOAD TADPOLES
Often called "toadpoles", these are jet black and chunky, and gather in large groups on the surface of the water. They eat algae but develop a taste for meat as their legs develop.

Plants that lay dormant over winter at the bottom of the pond, such as water soldiers, hornwort and water lilies, gradually rise to the surface in mid-spring, and yellow flag iris and marsh marigold, the first of the marginals, start flowering, providing food for bees and other pollinators.

Around the pond, plants begin to grow strongly, and moths, leafminers, aphids and froghoppers lay eggs on their leaves. You may spot house sparrows taking aphids and caterpillars from plants to feed their young. Even the mud around ponds is put to good use by swallows and house martins building their nest cups, while some bee and wasp species take small amounts of mud to seal their solitary nests.

During a dry spring, a pond becomes a lifeline for huge numbers of wildlife, which rely on water to stay alive. You might spot honey bees resting on a lily pad to drink from the surface, or mammals coming to the edge to have a drink. Keep an eye out for the first bats of the year, which dart above ponds from dusk, feasting on midges, mosquitoes and moths.

Gathering mud
House martins gather mud from ponds to construct their nests just under the eaves of buildings.

Honey bee drinking
Ponds are a valuable source of water for honey bees, helping them to control temperatures in the hive.

In a dry spring, a garden pond becomes a lifeline for huge numbers of wildlife.

Plenty of aquatic life is breeding in mid-spring, from water beetles and backswimmers to mayflies and pond snails. Tadpoles have hatched and are starting to grow legs. At this stage they switch from being herbivores to omnivores, developing a taste for meat as they encounter dead insects on the water's surface, or even bits of dead snail or mammal they are lucky enough to happen upon. If there aren't enough food sources naturally in the pond, tadpoles may turn on one another. To stop them cannibalizing each other, try feeding them cold-water fish flakes to provide the extra food they need. Bear in mind that fish flakes contain a lot of nutrients so add a little at a time, and make sure you see the tadpoles eating them; otherwise, you could create the perfect conditions for algal blooms to develop.

Developing tadpole
In summer they develop legs and absorb their tails, leaving the pond as young adults.

SMOOTH NEWT
Lissotriton vulgaris

This common species of newt is fairly widespread across Europe, absent only from the Alps, southern France, the Iberian peninsula, Italy and most of the Mediterranean islands. Growing up to 10cm (4in) in length, it has a grey-brown body with a black-spotted orange belly. Males develop a smooth crest along their body and tail during breeding season.

Adults are found in ponds from late winter to midsummer. They eat invertebrates such as beetles, small slugs, and snails. In ponds, you may also see them eating frogspawn and preying on frog tadpoles. Outside breeding season, they are found on land, often beneath logs and stones, where they also hibernate.

BREEDING
A complicated affair, breeding takes place from late spring. The male swims around the female before wafting pheromones (chemical scents used to attract a mate) towards her with his tail. If she's interested, he will leave a packet of sperm (a spermatophore) for her to pick up with her cloaca (reproductive

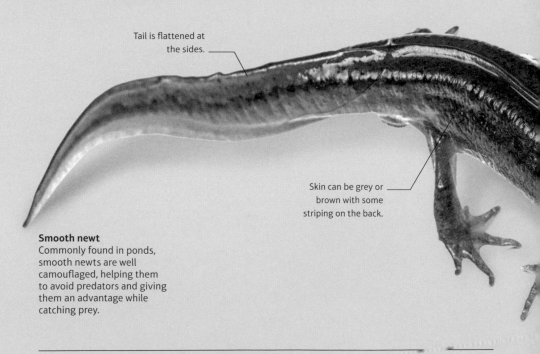

Tail is flattened at the sides.

Skin can be grey or brown with some striping on the back.

Smooth newt
Commonly found in ponds, smooth newts are well camouflaged, helping them to avoid predators and giving them an advantage while catching prey.

opening), and fertilize her eggs
internally herself. She then lays
her eggs individually, wrapping
each of them in a submerged leaf
of a marginal plant such as brooklime
or water forget-me-not. Each female
will lay up to 500 eggs, which she
fertilizes with sperm from a number
of different males.

Newt efts (juveniles) look like tiny
newts with long, feathery gills that
enable them to breathe underwater.
Unlike frogs and toads, they develop
their front legs before their back legs.
By late summer, at around three
months old, the young newts lose
their gills, at which stage they are
ready to emerge from the water and
live as adults.

NEWT EGG
The female lays eggs individually, wrapping them in
the leaves of marginal plants, from late spring to
early summer. She seals the folded leaf to make the
egg less vulnerable to predators.

A NEWT EFT
Newt efts have long, feathery gills near their head,
which allow them to breathe underwater. They stay
in the pond until late summer, by which time they
have absorbed their gills.

NEWT IN THE WATER
A wavy crest appears along the back of the male
newt during the breeding season. Both males and
females have a spotted belly and throat, which can
make it easier to see them in the water.

Towards the end of spring our ponds are lush and teeming with life. Plants around the pond grow tall, providing flowers and leaves for a range of insects including bees and other pollinators. Spring is a key time for pollinators. Many start to breed at this time and some, such as queen bumblebees, emerge from hibernation, having not eaten for several months. Nearly all pollinators drink nectar to give them the energy to fly and find a mate, while bees use protein-rich pollen to feed their young, and butterflies and other insects lay their eggs on leaves.

In the water, pond plants put on a lot of growth and quickly colonize the whole pond. You may want to remove some growth if the pond is looking overcrowded (see p.173), but bear in mind that many species will be using the pond at this time, so check thoroughly before disposing of it.

Tadpoles look more like their adult selves now, while nymphs are growing apace. Look into the water and you might see damselfly and dragonfly larvae or water beetles and tiny snails. On sunny days adult frogs may rest in the water, or newts rise to the surface to catch tadpoles. Honey bees and other insects drink from the safety of a water lily leaf or at the pond's edge.

Keep an eye on your dragonfly perch for the first dragonflies of the year. Many species pick a perch to rest on, which they return to in order to eat their food – usually flying insects. Regular night visitors may include bats and other mammals, which may be raising young in a nearby nest.

Night hunter
If you stay up after dark, you may spot bats flitting over the surface of your pond, hunting insects.

LARGE RED DAMSELFLY
Pyrrhosoma nymphula

A very common species, the large red damselfly is one of the first damselflies to emerge in spring, as early as March in some regions. Around 36mm (1½in) long, it can be found in a variety of freshwater habitats, including rivers, lakes, canals and garden ponds, where it hunts and snatches insects from vegetation.

Males are bright red with a black thorax and black bands at the end of the body. Females are more variable and less brightly coloured, with some appearing almost black.

Males aggressively defend their territory against other males, and occasionally other insects. You may spot them flying up from a perch to chase off interlopers, before returning when the coast is clear. Unusually for damselfies,

red damselfly nymphs are also territorial, although you're unlikely to see their territorial behaviour beneath the surface.

BREEDING
Mating takes place in late spring and summer. The male clasps the female by her neck and she bends her body around

Damselflies have beautiful chequered wings, which they close when they rest.

Large red damselfly
The male has a vivid red abdomen with black markings. Its legs are black and it has dark spots near the tips of its wings.

to his reproductive organs in an uncomfortable-looking formation called a "mating wheel". They remain paired and fly together over the water, with the female stopping to lay eggs on plants just below the surface.

The nymphs live in the water for up to two years, feeding on aquatic larvae. Eventually they climb out of the water, up emergent plant stems, and break out of their skin as adult damselflies. These live for a few weeks, just enough time to mate and lay eggs.

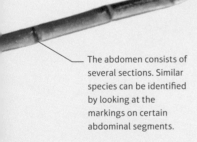

The abdomen consists of several sections. Similar species can be identified by looking at the markings on certain abdominal segments.

MATING WHEEL
The male perches on a plant while holding on to the female's head with claspers at the end of his abdomen. They can remain clasped together until the eggs are laid.

FEMALE DAMSELFLY
Females range in colour from red to almost black. Most damselflies hold their wings closed against their body when they are resting, unlike dragonflies which hold their wings out from their body.

SUMMER

It's pretty wild in my garden in summer. The pond is surrounded by wildflowers including water figwort and common valerian. I let these grow tall, and can barely see the pond from June to August, when I cut back the garden meadow and give the plants by the pond a good trim.

I always keep a bit of foliage short around one edge of the pond throughout this time so that I can see what's going on in the water. Frog and toad tadpoles are in the final stages of metamorphosis by July, and are usually starting to emerge from the pond. I wait with bated breath for the first urgent hops of tiny frogs into the long grass. I tread very carefully in the garden in midsummer.

Backswimmers seem to love my pond and breed in it every year. The adults are huge, imposing-looking insects, with strong jaws for piercing and sucking out the insides of their prey. I don't see them much – they spend most of their time at the bottom of the pond, coming to the surface only to trap air beneath their

Balanced planting
A mix of plants at the edges, underwater and floating on the surface offers food and shelter for all sorts of wildlife.

Cooling dip
In summer, frogs may rest in shade by the pond, or in the water with their heads above the surface, waiting to catch insects.

abdomen so they can breathe underwater. I love watching the backswimmer nymphs grow. They're feisty little things: I watch them chasing each other in circles just below the surface. I make sure I have as many meals by the pond as possible in summer, as long as the weather's nice. Just sitting next to the water helps clear my head, and there's always something new to spot, both in and out of the water. I might see a clear sausage of pond snail eggs on the underside of a submerged leaf, mating dragonflies and damselflies, or a honey bee drinking water from the surface. If I stay still for long enough, the birds come to the pond for a drink and a bathe – watching birds bring their fledglings to the pond for the first time makes me very happy indeed.

On really hot days frogs rest in the water. Slow worms bask in the woodpile next to the pond and insects buzz noisily. There's always some insect dipping its abdomen in the water to lay eggs. Sometimes I do a pond dip (see pp.96–99), using a fishing net and a couple of white baking trays, and marvel at the variety of critters that inhabit my pond.

At dusk I watch bats fly above the garden, catching mosquitoes and midges. I've recorded just one bat species here: the common pipistrelle, an abundant species that thrives in

Hedgehog foraging
While they are out hunting, hedgehogs may visit your pond for a drink.

urban areas like mine. I'm hopeful that, as the garden matures and more wildlife uses the pond, I might attract more species to the area.

I set trail cameras to watch hedgehogs at night. Sometimes I stay up to watch them, a blanket and a glass of wine to keep me warm while I wait to hear them snuffling in the borders. I think there are five individuals that regularly visit my garden, and two that sleep here during the day. At this time of year they make little tunnels through the foliage to reach the pond for a drink: little "desire paths" made through the border and meadow, all to and from my wonderful pond.

There's always
something new to see,
in and out of the water.

All the activity of spring continues into the early days of summer, as young amphibians are almost ready to leave the water for a new life mostly on land. It's a good idea to let grass and other plant material grow longer at this time, as they help shelter the baby frogs, toads and newts from garden predators such as birds and cats.

Creating log piles and similar habitats around your pond (see pp.24–27) can provide additional protection for these vulnerable young amphibians. These habitats also provide cover for the insects and other invertebrates amphibians eat, giving them lunch and a home in one.

Plant life
Pond plants grow
and expand quickly
in early summer.

You might spot newly laid
invertebrate eggs in the
water, or larvae or nymphs
swimming around in the
shallows. The predatory
aquatic larvae of beetles,
bugs and dragonflies hunt tadpoles in the water, and catch
drowning insects such as flies and bees on the surface.

Plants around the pond are bulking up, providing food for
a huge range of wildlife. This is also a crucial time for small
mammals and birds, many of which will be nesting nearby,
and need to use your pond for drinking and, to a lesser extent,
bathing. You may see birds taking insects and caterpillars to
feed their young, or bees and other pollinators visiting flowers.
Small mammals such as mice and voles may shelter among
any long grass around your pond, and you may be lucky enough
to have grass snakes hunting there. At night look out for moths,
which will not only visit flowers but may also lay eggs among
the grass and foliage. Bats, too, benefit from plants growing
around ponds, as plants provide them with more insects
to feast on.

Natural food
Blue tit chicks need to eat
a huge number of larvae
and caterpillars in the first
few weeks of their life.

GREAT DIVING BEETLE
Dytiscus marginalis

This large, olive-brown beetle is widespread in Europe and can be commonly found in ponds, lakes, and canals. It is about 3cm (1¼in) long, so is easy to spot, but you are most likely to see it as it comes to the surface of your pond to replenish its air supply, which it stores beneath its wing cases.

It eats small aquatic invertebrates, tadpoles, and even small fish. Adults fly

Powerful jaws enable beetles to prey on a huge range of aquatic invertebrates, including small fish.

Ribs on the wing cases of females help the males cling on to them during mating.

A yellow border runs around the side of the beetle, from the thorax to the wing cases.

between ponds at night, guided by the reflection of the moon on the water's surface. Sometimes (presumably confused by the moon's reflection) they land on car bonnets and in puddles on the road.

GREAT DIVING BEETLE LARVA
You may be lucky enough to have great diving beetle larvae hunting in your pond, like this one here, predating a tadpole. The larvae can grow up to 5cm (2in) long.

BREEDING

Mating takes place in the water. Males have suction pads on their front feet in order to grip the females when mating, and females have ribbed wing cases to make them easier to cling on to. Eggs are laid in the leaves of pond plants.

Great diving beetle larvae are voracious predators, just like the adults, attacking water invertebrates with their large, biting jaws. They pupate in damp soil by the edge of the water, ready for a lifetime of more aggressive, predatory behaviour – hopefully in your pond!

COMING UP FOR AIR
Sit patiently at the water's edge and you may spot an adult great diving beetle coming to the surface to top up its air supply. It does this by poking the tip of its abdomen up through the water surface.

Predatory beetle
This large beetle is a fearsome predator as a larva and as an adult. The female can be recognized by her ribbed wing cases.

MALE GREAT DIVING BEETLE
Smooth, shiny wing cases distinguish the male from the female. The front two pairs of legs also have suction pads so they can grip the female while mating, and hang on to prey.

As temperatures rise, dragonflies and damselflies make
the most of fine weather by laying eggs among submerged
pond plants. The males often stay with their mates while
they are laying eggs, either linked or close by. These larvae
are welcome predators of mosquito and midge larvae, while
bats do a good job of controlling numbers of adults in turn.

Now is the best time for pond dipping (see pp.96–99) as
the pond is so full of new life. You'll catch anything from water
beetles and diving beetles to water fleas, dragonfly nymphs,
caddisfly larvae and pond snails.

By now most of the year's young amphibians have left the
water, and may be found sheltering from predators among
low-growing plants, long grass and log piles. They will spend the
rest of the summer eating small insects and other invertebrates,
growing to adult size before overwintering among fallen leaves,
logs, other plant debris, and compost heaps. They won't be
sexually active for two or three years.

Pond water levels often fall in summer, especially in very dry
weather. Natural fluctuations in the pond level are normal and
nothing to worry about as pond animals are well adapted to
such changes, so avoid topping it up unless the water level gets
so low that mammals are unable to get in and out safely. If you
must top up your pond, ideally use rain water rather than tap
water, as tap water can contain chemicals that harm pond life
(see p.178). Waiting for late summer rains to replenish the water
in the pond is often the best policy.

Laying eggs
The female dragonfly lays
eggs just under the water, on
to submerged pond plants.

GRASS SNAKE
Natrix helvetica

Sometimes found in gardens and allotments – particularly those with ponds and compost heaps – grass snakes are completely harmless. Non-venomous and timid, they spend most of their lives sheltering beneath logs and stones, emerging only to mate and hunt prey.

Adults can grow up to 1.5m (5ft) long, and are olive green with a yellow collar. They eat amphibians, fish, birds, and small mammals. Juveniles eat small invertebrates such as froglets and slugs. *Natrix helvetica* is found in England and Wales, as well as Italy, France, parts of Germany and Switzerland. No grass snakes are recorded in Scotland or Ireland.

In summer, they bask in the sun near ponds, usually beneath logs or bespoke "reptile tins" which you can lay out over long grass. If you're lucky, you may even spot them swimming across the surface of your pond in search of prey.

If you accidentally disturb a grass snake, don't be alarmed if it produces a garlicky-smelling excretion or feigns death. Without a venomous bite, it has evolved these measures to put off predators, despite being completely harmless to humans.

Females tend to grow longer than males.

Excellent swimmer
Grass snakes swim well and can hunt in ponds, but also like to shelter beside the water.

BREEDING

After mating, females lay up to 40 white eggs in rotting vegetation, such as compost heaps, and piles of grass clippings, the warmth from which incubates them until they hatch in early autumn.

From October to April, grass snakes hibernate, choosing spots such as compost heaps, log piles and even old rabbit warrens to spend the winter.

REPTILE TINS

A reptile tin is a piece of corrugated tin (or slate roof tile) that offers shelter and warmth by soaking up the sun's rays. If you lay a reptile tin on long grass you may find grass snakes sheltering beneath it.

GRASS SNAKE EGGS

If you know grass snakes are in your area, be careful when adding or removing waste from your compost heap – you may be fortunate enough to find their eggs, around 25mm (1in) in length, in there.

The yellow collar is unmistakable, enabling you to tell grass snakes apart from other snakes and slow worms.

CLOSE COUSINS

Natrix helvetica, the species commonly found in England and Wales, was classified as a separate species in 2017. Before then it was thought to be *Natrix natrix*, which is widespread across Europe.

Activity in and around the pond starts to wind down by the end of summer; compared to early spring, everything is quiet. Aquatic larvae and nymphs are still present, but tend to stick to the bottom of the pond, often hunting smaller species and trying to avoid being eaten by those larger than themselves.

If you spot tadpoles in the water at this time of year, they could be suffering from "Peter Pan" syndrome. This is when tadpoles keep growing but never develop legs. It's more likely to happen in cold summers, with metamorphosis taking place more slowly than in warm conditions. Sometimes these huge tadpoles spend winter in the pond. Don't worry too much – tadpoles can survive winter in the water. This can even put them at an advantage in spring as they finish metamorphosis before other tadpoles, giving them more time out of the water to feed up and become a good size before their second winter.

"Peter Pan" tadpoles
Large tadpoles without legs can go through the winter and finish developing the following spring.

Unclogging ponds
Where plants are dying back or taking over the surface, remove some of them carefully to let light in and improve space around the plants.

With luck, late summer rains and storms restore pond levels to their spring capacity, breathing new life into the garden. Now's the time to give the pond a bit of a tidy – remove up to half of all plants growing on the surface, including algae and pond weed (see pp.172–173), and give plants at the edges a trim. Although most amphibians will have left the water by this point, you should still swill plants in a bucket of pond water before discarding, to dislodge any larvae, nymphs or even adult frogs that might be caught in the material. Return the water to the pond and add cut-off plant material to the compost heap.

By the end of summer, everything is relatively quiet as activity starts to wind down.

COMMON DARTER DRAGONFLY
Sympetrum striolatum

One of the most widespread dragonflies across Europe, the common darter dragonfly is found in a number of habitats from late spring to late autumn, including along grassy woodland rides and canals, as well as around garden ponds. They can often be found a good distance from water, either searching for a mate or hunting prey.

Males are red with yellow-striped legs, and females are yellow-green, ageing to a dull brown-red. You might spot them basking in the sun to warm up in the morning, always with their wings at right

Wing spots may vary in colour.

Like all dragonflies, the common darter rests with its wings outstretched, unlike damselflies, which close their wings.

They use their front legs as a "basket" to scoop up prey, returning to their perch to eat it.

Darting predator
This dragonfly darts out to catch its prey (usually an insect), taking it back to a perch to eat.

DRAGONFLY NYMPH
Living in the water and dull in colour, dragonfly nymphs are voracious predators of other pond life, including tadpoles. They have six legs, like adult dragonflies, and strong jaws.

angles to their body, unlike damselflies, which rest with their wings tucked in. As their name suggests, they "dart" to catch insects such as midges and mosquitoes, then return to a favourite perch to eat them – if you place a "dragonfly perch" near your pond you will almost certainly encourage dragonflies to rest on it (see pp.24–25).

BREEDING

Mating takes place in summer. The male and female form a "mating wheel", where the male clasps the female by her neck and she brings her abdomen to his body to fertilize her eggs. She lays eggs among submerged pond plants just below the water's surface, often with the male flying in tow. Eggs laid in summer hatch within a couple of weeks, while those laid later in the year hatch the following spring. The nymphs are predatory, eating other aquatic invertebrates and larvae, including water fleas, snails and tadpoles. They emerge a year later, between June and October, by climbing up a stem and breaking out of their final nymphal case as adult dragonflies.

EMERGING DRAGONFLY
Look closely at tall stems of pond plants and you may spot dragonfly nymphs emerging from the pond for their final transition into adulthood. They wait until their wings harden before flying away.

MATING WHEEL
Like the damselfly, the common darter male and female join into a mating wheel to fertilize the female's eggs. The male will often fly with the female while she lays her eggs.

AUTUMN

The pond is quieter in autumn. All frog
and toad tadpoles have left by now, while
other aquatic nymphs and larvae – such
as backswimmmers, hoverflies and other
flies – have reached their adult stage.
I peer into the water and see nothing.

I don't know where the baby frogs and toads go in
autumn. I assume they settle down under the log piles
or among the compost in the large heap at the back.
I see them only in late summer when they leave the
pond, the meadow alive with little bodies, crawling or
hopping their way to safety. I hope they find somewhere
nice to hibernate in the garden. Some adult frogs will
undoubtedly overwinter in the mud at the bottom
of the pond. I keep an eye out for them swimming
across the bottom on sunny winter days, but I've not
been so lucky as to see one.

   I cut the meadow in early autumn, so I can see the
pond fully from my back door again. I see birds bathing
for the first time in weeks. I have to cut the meadow
very carefully to avoid harming wildlife, and so I do it in
stages, leaving some areas of long grass until October.
I start the cut by using a broom to disturb and check
the grass, transferring any critters I find along the way
to the log piles on either side of the garden. Then I

Autumnal pond
The pond becomes
quiet as plants die down
and animals hibernate.

Regular bathers
House sparrows are often among the most frequent pond visitors.

use shears to cut the top half of the meadow before checking again and strimming the rest. It sounds labour intensive, but I have a small garden and it means I don't harm anything living among it. Also, I usually find exciting things among the grass and wildflowers that I wouldn't normally see – big fat caterpillars, the odd mouse, a toad.

There's a patch of meadow at the side of the pond that I don't cut at all. I use this as a "buffer" for anything that isn't yet ready to leave the pond, or for those that have, but still need some shelter. It seems cruel to scalp the whole thing and leave the wildlife so exposed. I like to watch the flowers and grasses degrade slowly over the autumn, before frost finally finishes them off in winter – by spring you can barely tell I didn't cut it.

Around the garden I move plants that were growing in the wrong place, avoiding the big autumn tidy up so many gardeners undertake at this time of year. I leave seedheads standing and plants to rot into themselves – any and all cover is used by some wildlife. I mulch the ground with leaf mould,

I leave seedheads
standing – any and
all cover is used by
some wildlife.

a wonderful material made from the previous year's autumn leaves, which I collect and bag up behind my shed. This stuff is full of beetles and other invertebrates that amphibians and birds love to eat. I watch blackbirds pick through it for food, and think of frogs and toads perhaps hiding beneath it, snacking on the odd beetle here and there.

The one area I do tidy is the pond itself. By autumn the brooklime and water forget-me-not have completely covered the pond, so I cut it back so that at least one-third of the pond's surface is clear. I remove as much algae and duckweed as I can, too, along with any pond detritus such as fallen leaves or plant stems that might increase nutrient levels as they break down, which would lead to more algae developing (see pp.172–173). All plant material and algae goes into a bucket of water for a few days before I compost it, so I can be sure no wildlife ends up in the compost bin.

Autumn maintenance
Autumn is a good time to cut back pond plants and remove pond weed.

The pond clings to the last of its summer beauty for another few weeks. Water lilies may still be flowering and plant material still growing. Yet, as invertebrates complete their lifecycle, the pond empties of fun. No more are the midge larvae dancing on the water's surface or backswimmer nymphs chasing each other beneath them. Tadpoles are but a distant dream and even the young amphibians are nowhere to be seen.

Around the pond you may have long grass and wildflowers, which will be seeding now. Look out for seed-eating birds, which cling on to thin stalks and eat from the seedheads as they bend to the ground with the weight of them. You may even spot mice and voles feasting on the seeds, climbing the stems and chewing the seedheads off so they fall to the ground for easier eating.

Eating seeds
Small mammals, such as this bank vole, make short work of autumn seeds.

Late flowers
Water lilies may be
flowering their last
in early autumn.

Like all plants in pots, pond plants in planting baskets need repotting every so often. This is a good time to do this, while water temperatures are mild (see p.173); remember to take care not to disturb any wildlife hiding among the foliage. If the plants also need dividing, wait until spring when active growth begins (see p.174).

Logs, large stones, open compost heaps and tiny crevices behind fence posts and beneath sheds become homes for a range of pond life. These out-of-the-way places provide the perfect quiet spot to hibernate, where anything from newts to pond skaters will hopefully remain undisturbed. Amphibians will seek out slightly damp habitats while insects will look for drier shelter. On mild days you may still spot adult frogs in and around the pond. Look out for great, fat females, already full of spawn. They carry this throughout winter, ready for breeding in spring.

Tadpoles are a distant
dream – the pond
empties of fun.

COMMON PIPISTRELLE
Pipistrellus pipistrellus

This tiny bat is widespread across Europe, and is the most common bat species in the British Isles. It's found in rural and urban areas and is the bat you're most likely to see flying over your garden. It has a fast, jerky flight. Like most bats it eats insects, which it catches using echolocation (a series of shouts which bounce off their prey and return echoes) so it can gauge how close it is before swooping in for the kill. If you want to hear bats you'll need a bat detector as most bats echolocate above the range of human hearing (up to 20 kHz depending on age). The common pipistrelle echolocates at 45 kHz and the related soprano pipistrelle (named as a separate species only in 1999) at 55 kHz. Once you've set your bat detector, you can "hear" the bats echolocating around you.

All bats have their own favourite food and the common pipistrelle eats small

Pipistrelle hunting
Ponds are a valuable source of insects for pipistrelles, which flit over the water when hunting.

Bats' wings are made of two layers of skin, blood vessels, and muscle fibres, which cover the wing skeleton.

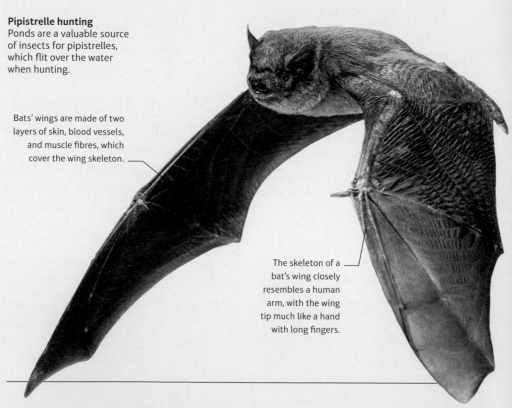

The skeleton of a bat's wing closely resembles a human arm, with the wing tip much like a hand with long fingers.

flies, mosquitoes and midges. It can catch up to 3,000 insects a night, which it eats "on the wing".

BAT BOX

Natural nesting sites can be limited, particularly in urban areas, so a ready-made bat box provides valuable shelter. Your pond encourages bats to feed in your garden; a bat box encourages them to stay.

BREEDING AND BAT ROOTS

Common pipistrelles roost in different locations depending on the time of year, often communally in crevices such as beneath house roof tiles, in tree holes and in bat boxes. Females form maternity colonies in summer and give birth to one (occasionally two) "pups" in June or July. They feed them milk for three to four weeks. Then the young bats are able to fly and after six weeks they can forage by themselves.

EXCELLENT EARS

Bats' ears are large in proportion to their body. They detect insect prey by listening to their own calls rebounding from the insect's body. Bats also make social noises to communicate with other bats.

Shortly after weaning, mating territories are set up for the following year's young, often near a winter roost. Males patrol these territories while "singing" to females as they fly to the roost. These territories last from July to October, with most mating taking place in September. The females store the male's sperm until the following spring, when they fertilize their eggs.

BAT DIET

Most bat species catch and eat invertebrates such as moths and midges on the wing, but some also take prey from leaves and bark, and find it easier to eat large prey while they are still.

As autumn progresses, the garden gradually shrinks into itself. Seeds fall to the ground and leaves and stems turn brown and lose their vibrancy. Conversely, the pond might look bigger as plants growing around the edge retreat, revealing more water.

Water lilies die back to their basket in the deeper parts of the pond. If you planted them recently, make sure that the pot sits deeply, well below where ice may form. Oxygenating and floating plants sink to the bottom of the pond for winter. You may think they have died or disappeared, but wait until mid-spring before replacing them, as they will most likely resurface.

Any long grass near the pond can be cut now (see pp.173–174), but check it first for wildlife. Many species, such as pond skaters and water beetles, hibernate around the pond edge in winter. Take extra caution when cutting plant material back, as you could unwittingly expose some insects, making them more likely to be predated. Check material well before composting it, and add it to a large open heap, so that some insects can return to shelter if they need to. You may want to leave a buffer for hibernating insects and late amphibians, as I do.

If you haven't already, now's a good time to give the pond a clear out. Remove debris such as fallen leaves and old plant stems (see p.172) from the bottom to stop the pond silting up, and cut back plants so they're covering no more than two-thirds of the surface. Don't go overboard when clearing out the pond, however. The silt and mud at the bottom is vital for hibernating invertebrates and amphibians.

Shelter at the pond edge
Leave some plant stems standing, such as those of common rush (*Juncus effusus*, pictured). They will provide shelter for hibernating insects until spring.

HEDGEHOG
Erinaceus europaeus

The European hedgehog is one of 17 species globally, and Britain and Ireland's only spiny mammal. Rounded and brown, it has yellow-tipped spines over its back and a furry face and "underskirt". As they are mostly nocturnal, hedgehogs may be seen snuffling around the garden on summer nights. Look out for black droppings, about 5cm (2in) in size, often containing shiny beetle wing cases. These are a good way to tell if hedgehogs are regularly visiting your garden. Although commonly associated with eating slugs, hedgehogs actually eat a wide variety of invertebrates, particularly beetles, caterpillars and earthworms. In fact, it's thought that they mostly avoid slugs, which can give them lungworm. Hedgehogs can also eat amphibians and birds' eggs. The best extra food we can offer them is cat and dog food – particularly kitten biscuits.

BREEDING
Mating takes place from late spring to late summer. The male circles around the

Hedgehog spines are modified hairs. Each has a hollow shaft and a muscle so the hedgehog can raise it defensively when it feels threatened.

A hedgehog's face and "underskirt" are covered in coarse hairs.

female, sometimes for hours, trying to persuade her to mate, which involves her flattening her spines to avoid harming him! Pregnancy lasts for four to five weeks, depending on the weather, and litters can contain up to seven hoglets. Born blind and without hair or spines, they feed on their mother's milk for the first few weeks, before being able to join her on foraging trips. After another couple of weeks the hoglets can live independently and will become solitary.

In warm regions, females mate again and have a second brood, sometimes as late as September. These "autumn orphans" can arrive too late in the year to be able to fatten up in time for winter hibernation and often don't make it through winter.

Hibernation takes place from late autumn to early spring. As the weather turns, hedgehogs will eat as much as possible to gain the fat reserves they need to survive winter. They hibernate beneath hedges, in large log piles or compost heaps, adding dry leaves and other materials to keep them snug.

Hedgehog
Ponds are a safe source of water for hedgehogs if there is a ramp or pile of stones the hedgehog can use to access the water.

ACCESS TO WATER
Ponds and shallow water sources are vital for hedgehogs in a dry year. Hedgehogs are vulnerable to dehydration, so safe and local water sources are essential for their survival.

HEDGEHOG OUT IN THE DAY
It's not unusual for hedgehogs to wake up from their hibernation periodically during winter, but if you see one out during the day you must take it in and call your local hedgehog rescue.

HEDGEHOG HIBERNATION
Hedgehogs hibernate typically from October to March, deep in leaf piles, beneath hedges and compost heaps. Take care when gardening in autumn and winter, to avoid disturbing them.

Your pond should be close to dormancy by now, with marginal plants trimmed back and debris removed. Birds will still bathe and mammals will still stop by for a drink, but those that use ponds to live and breed will be very quiet now, as the water starts to cool and breeding season is past. Amphibians should be hibernating now, although on mild days you may hear the odd croak of a male frog, eager for winter to be over and done with, perhaps.

Some aquatic invertebrates, such as pond skaters, hibernate away from the pond, often in groups or clusters. Others, such as water beetles, will spend winter in the mud at the bottom, rising occasionally to replenish air stocks. Frogs hibernate in a variety of locations, such as beneath compost heaps and log piles, but some will rest in the mud at the bottom of the pond, breathing through their skin. On sunny days you may spot one swimming across the bottom.

Drinking spot
Mammals such as squirrels may stop regularly at your pond to drink.

Hibernating wasp
Queen wasps hibernate in tree cavities or buildings. They emerge the following spring to start a new nest.

Around the garden, insects and other invertebrates will be hunkering down. Everything in your garden has the potential to be a hibernaculum (place to shelter wildlife). Avoid turning your compost heap as it could be home to anything from bumblebee queens and ground beetles to mice, voles and hedgehogs. Slow worms bury themselves into the soil, adult butterflies tuck themselves behind climbing plants or beneath shed roofs, while queen wasps hide behind thick bark.

Use autumn leaves to make leaf piles for hibernating insects and amphibians, ensuring they won't have far to travel to return to your pond in spring. Tuck them beneath a hedge or behind your shed, or use chicken wire to make a bespoke "leaf cage", leaving a gap at the bottom for amphibians to squeeze beneath. Providing safe places for wildlife to hibernate will give you a much richer, noisier garden next year.

Use autumn leaves to make leaf piles for hibernating insects.

COMMON BACKSWIMMER
Notonecta glauca

Also called water boatman, the common backswimmer is a large aquatic invertebrate, found throughout Europe and into north Africa and Asia. It grows up to 2cm (¾in) in length and swims on its back, as its name suggests. It has long, flat, oar-like hind legs, adapted for swimming, and breathes by trapping air among hair-like structures all over the body, which creates a film visible to us as a silvery sheen. This film not only enables it to breathe but also keeps its body dry while underwater. Adults fly between habitats and can be some of the first species to colonize a new pond.

It's found in a variety of watery habits, including ponds, water butts and cattle troughs. A voracious hunter, it preys on insects on the surface of the water, along with tadpoles and small fish, catching them with its strong, short forelegs and sucking out their insides using its beak-like rostrum. Take care when pond dipping or clearing out your pond – it can deliver a painful bite.

BREEDING

Mating takes place in spring, with oblong eggs laid among submerged pond plants. The eggs hatch into nymphs, which go through "incomplete metamorphosis"

If you catch a backswimmer while pond dipping, you should be able to see air trapped in hairs around the body, visible as a silvery sheen.

and therefore resemble the adults. The nymphs have much shorter abdomens than the adults and less developed wings. They go through five nymphal stages over the summer, reaching adulthood in autumn. They tend to remain in the pond all winter.

BACKSWIMMER NYMPH
Like the adult backswimmer, the nymph is a very effective predator. It develops its full-sized wings at its last nymphal stage before becoming an adult. Its large eyes have excellent vision.

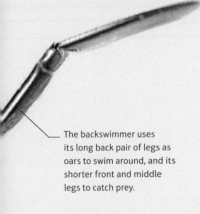

The backswimmer uses its long back pair of legs as oars to swim around, and its shorter front and middle legs to catch prey.

HUNTING INSECTS
Backswimmers can dart incredibly quickly when they detect movement at the water's surface, catching their prey with their legs and quickly piercing it with their mouth parts.

Swimming upside-down
Backswimmers paddle through the water grabbing insects on the surface. With their distinctive, fast glide, they are easy to spot in the pond.

BREATHING AT THE SURFACE
Although it spends its time underwater, the backswimmer breathes air at the surface, using the air to surround its body in a film which keeps it dry and supplies oxygen.

WINTER

I like the pond in winter. It's so quiet and
still at this time of year, and there's a
peacefulness to it that radiates out
across the garden.

I keep the bird feeders filled with sunflower hearts and
suet nibbles. I top up the bird baths, despite the birds'
apparent preference for the pond, where they continue
to bathe daily. House sparrows descend on the pond en
masse, squeezing as many of themselves as possible on
to the dragonfly perch, before dunking themselves into
the water and emerging a few seconds later, looking like
drowned rats. The blackbirds bathe at the other end, in
turns, taking their time to dunk and ruffle. The winter
robin darts in and out, while the goldfinches still seem
too busy taking the last of the knapweed seedheads to
bother with such trivialities as washing.

Bird life in my garden becomes more exciting in
winter. Tits and finches gather in large roving groups,
looking for food. Whereas one or two great tits or blue
tits might turn up in spring, I might see 10 at a time in
winter. Likewise the goldfinches, which already gather in
loose family groups, are joined by 20 or so greenfinches.
These big, green birds have suffered huge declines in

Pond in winter
Snow and ice give a
different, more structural
beauty to a pond and the
winter stems around it.

Hungry great tits
You may spot roving
gangs of tits in winter,
which tend to search
for food together
before dispersing
again in spring.

recent years thanks to trichomonosis, a parasitic disease that
affects their ability to eat. I love seeing them in my garden,
knowing they can breed and shelter safely – I clean my bird
feeders regularly so diseases and parasites don't build up.

I watch for migrant redwings and fieldfares, which fly here
from Scandinavia for winter. Any bird can turn up in any garden
in winter – woodcocks from Finland and Russia often turn up in
London gardens, exhausted by their journey and en route
elsewhere. Other birds, such as bramblings and redpolls, appear
in gardens when natural sources of food are in short supply. It's
always worth looking up, and looking in the garden, in winter.

Hedgehogs are hibernating now, although there's always one
that turns up occasionally, for a drink in the pond and a snack on

cat biscuits. It seems healthy, and it's so mild in Brighton that hedgehogs here rarely hibernate fully from November to March. I make sure I leave food out until it's no longer taken and keep the trail cameras on for signs of unhealthy or underweight hogs that need a hand getting through winter.

Sometimes, but sadly only very rarely, the pond ices over. I go out and check after a heavy frost, tracing patterns in the ice as it formed over leaves and other obstacles, with my fingers. We never have a prolonged cold spell here in Brighton, but a bit of ice at some point over winter helps the feeling that the world is still turning, that the basic tenets of seasonal change are still in place, despite being early or late, short-lived or barely visible.

Towards the end of winter, I get impatient for spring. There may be ice and frost, but there are also mild days and rain, and with them come frogs. I scan Twitter for the first cries of "frogspawn!" from Devon and Cornwall, and trace sightings in the south west until I know it's only a matter of days before I spot my first clump. Winter is the dreariest season, but in my head half of it is spring, or at least preparation for spring. I sit by the pond for a glimpse of movement, and set trail cameras on the back gate to see amphibians arriving for their giant mating party. This, the end of winter, is my favourite time of year, because so many wonderful things are just around the corner.

Redwing
From the thrush family, redwings can be seen in flocks in winter, and feed on berries or worms.

Early winter is a time to focus on birds. All birds need water for drinking and bathing so your pond will already be a rich resource for them. If you leave seedheads on plants around your pond you may spot goldfinches and sparrows feasting on the seeds. On mild days you may see blackbirds and robins picking through plants at the water's edge for worms, grubs and insects. Berrying shrubs and trees provide additional sustenance.

When the ground freezes, it's important to give birds an extra helping hand as they are unable to forage for natural food. Put out halved apples, seeds, nuts and small amounts of grated cheese and crumbs. If your pond freezes over, leave a dish of fresh water daily, so birds can continue drinking and bathing.

This is a great time to keep an eye out for migrants. Winter migrants fly from colder to warmer regions, usually travelling in a southerly or westerly direction. In the UK scores of redwings, fieldfares, waxwings and hawfinches arrive from Russia and Scandinavia – sometimes in huge numbers. Other birds, such as

Extra bird food
For blackbirds and other ground feeders, cold, hard ground is harder to forage from. Leaving out additional food such as halved apples can be a lifeline for them.

Visiting fox
Foxes stay in a smaller area over winter and may visit your garden more often.

robins and blackbirds, migrate from, say, Sheffield to Cornwall or Brighton to Brittany. Never count on that robin you see every day in your garden being the same one 365 days a year – it's much more likely to be two individuals, one for spring and summer and the other for autumn and winter.

Migrant birds may not be as used to humans as resident birds are, so could be too timid to visit busier areas of the garden. If your pond is exposed, consider leaving a dish of water somewhere quiet and out of the way, along with some food so they can eat and drink under cover.

While foxes don't hibernate and may continue visiting your garden, look out for hedgehogs, which should be hibernating. They may move between hibernation sites on mild nights but should never be out during the day (see p.147).

Keep an eye out for migrant bird species coming to your garden.

BLACKBIRD
Turdus merula

The blackbird is a gorgeous medium-sized bird, common in Europe and west Asia. The male is glossy black with orange eye-rings and an orange bill, while the female is dull brown, with thrush-like chest markings. It eats invertebrates, particularly earthworms and caterpillars, and fruit such as fallen apples, cherries and blackberries.

Males can be heard singing from rooftops from March throughout summer. They have a glorious, melodious, flutey song. Later in the summer young males can be heard practising their song for the following breeding season. Listen out for low, melodious gurgling from deep within a hedge or shrub. This song is called "sub-song" and sounds a bit like the bird is trying to sing quietly, under his breath.

BREEDING

Depending on the weather, breeding season takes place from March to late July. The female builds a nest of twigs and grass, usually in a bush or a small

The orange eye ring and bill don't develop in male blackbirds until the second year.

Blackbirds forage on the ground for insects, fruit and caterpillars, turning over leaves with their feet.

tree, close to the ground. After mating she lays three to five blue speckled eggs, which she incubates for two weeks, until they hatch. Both parents feed the chicks worms and caterpillars. The young are ready to fledge (leave the nest) after two weeks, but can't fly for another week after fledging. They're fully independent three weeks after leaving the nest.

Fledged chicks are often left in the care of the male while the female prepares for the next brood. Care of the third and final brood of the season is usually divided between the parents, with each adult taking sole care of some of the young.

While three broods of up to five chicks may seem like a lot of baby blackbirds, predation is high, with only 30–40 per cent of nests bearing fledged young. Cats, magpies, stoats, weasels and even squirrels will take blackbird chicks, both from the nest and once they have fledged.

Male blackbird
Medium-sized, and glossy black in colour, the male blackbird is a distinctive garden visitor all year round.

FEMALE BLACKBIRD
Similar in size to the male, the female is brown with a speckled breast. She is usually the main nest builder and may have up to three broods of chicks. She shares feeding of the chicks with the male.

A CLUTCH OF BLACKBIRD EGGS
The blackbird's nest is an intricate cup of grass lined with mud. It can take up to two weeks to build. Eggs are blue and speckled, which is thought to provide protection from the sun's UV rays.

BLACKBIRD FEEDING A CHICK
Blackbirds often feed chicks until they can forage for themselves. Look out for parents and their chicks visiting your garden, with chicks begging their parents for food.

This is the most sombre time of year for garden ponds. Many will be frozen over, hiding hibernating amphibians and pond plants beneath the surface. Above the surface, birds are the only visible signs of life, each one battling with short, cold days to find the energy it needs to stay warm at night. Keep bird feeders topped up and refresh bird baths and other water dishes daily. Continue feeding any mammals that may stop by.

When a pond freezes over, usually only the first few centimetres turn to ice. Ponds with deeper areas – up to 60cm (24in) – rarely completely freeze, enabling amphibians and other wildlife to hibernate safely beneath the surface. On the whole, those that choose to hibernate in ponds have evolved to survive temporary freezing conditions.

While we no longer tend to experience prolonged periods of very low temperatures, problems may occur if your pond freezes for a period of several weeks. It is thought that thick ice on the surface of the pond, with a lot of rotting vegetation underneath, prevents harmful gases from leaving the pond and can contribute to a phenomenon known as "winterkill" of amphibians (see p.179). Although algae can continue to photosynthesize (and therefore produce oxygen) under ice, if you are concerned, float a ball on the surface to stop it freezing completely. This will ensure mammals can drink from the water, too.

Bird feeders
Feeding birds in winter provides them with a reliable source of nutrition, while giving you the opportunity to view them up close.

Frozen surface
Algae can photosynthesize through thin layers of ice, but are not able to do so after heavy snow.

HOUSE SPARROW
Passer domesticus

Small and brown with a distinctive, loud "cheep", the house sparrow is very common throughout the world, although it is suffering declines in the UK, France, Germany and other parts of Europe, particularly in urban areas. The male has a grey cap and black bib that extends down onto the chest, and a chestnut brown back streaked with black and white. The female is sandy brown with brown and grey streaks on the back and wings. Juveniles of both sexes resemble the female.

House sparrows thrive in scrubby habitats, gardens, parks and woodland, particularly where hedges are present. Adults eat a variety of seeds, but feed invertebrates such as aphids, caterpillars and beetles to their young.

BREEDING

Adults live in loose family colonies in which they nest and feed together. Each pair nests from April to August, usually in

A sparrow's short, thick beak is used to crack open seeds and nuts.

Sandy colouring distinguishes the female sparrow from the male.

holes in houses and roofs, but also nest boxes and occasionally in hedges. The nest is made from a loose assembly of dry grass or straw and lined with feathers and other soft material. Pairs often remain faithful to their nest site and each other, and have up to three broods a year.

The official nesting season is from April to August, although nesting has been recorded throughout the year in mild regions.

The female lays up to five eggs. Both sexes incubate the eggs, and also feed the chicks, which hatch some two weeks later. The young fledge after a further two weeks and continue to be fed by the parents for around a week after leaving the nest, remaining with their parents for a fortnight. Often this parental duty falls to the male as the female prepares to lay the second brood.

FEEDING A FLEDGLING
It's obvious when house sparrows fledge, as you start to see nagging chicks and parents taking seed from feeders to pass on to them. Here a male, with its distinctive grey head and dark bib, feeds a chick.

TAKING A DRINK
Like all birds, house sparrows need water for drinking and bathing, and will use a shallow area of your pond for this purpose. They often bathe in large, noisy groups.

DUST BATH
Sparrows often take a dust bath after a water bath. They roll around in dry soil or dust, then preen out the dust. It's thought that they might be removing parasites at the same time.

Female sparrow
The female is overall sandy brown in colour with its head the same colour as the body.

Depending on where you live, your pond may start to show signs of life already. You may spot a frog resting at the water's edge, signalling that spring is on its way. Birds will continue to bathe and drink from the water, while hibernating mammals, such as hedgehogs, may start to rouse from their torpor, popping into your garden for a drink from your pond.

Now's the time to undertake any last-minute pond-clearing activities before amphibians and aquatic invertebrates start breeding. It's also a good time to add barley straw to the water, as an organic way to reduce algae (see p.177). As the barley straw breaks down it releases chemical compounds into the water, which inhibit algae from growing. This is best used in ponds with moving water and takes around a month to start working, depending on the temperature. If you add it to your pond in winter, it can start decomposing before the main spurt of algae growth starts in spring.

Bathing blue tit
Blue tits will use the pond for bathing and drinking all year round.

Male and female frogs
You may spot the smaller males hitching a ride on the back of the females en route to a breeding pond.

In mild winters, amphibians may already be on the move to their breeding grounds. Keep an eye out for frogs or toads travelling together, with males hitching a ride on the females along the way. Now's a good time to ensure amphibians can enter and exit your garden easily; cut a 13cm (5in) square hole at the bottom of your fence or garden gate, or dig a hole beneath it. Consider chatting to your neighbours to encourage them to do the same to create a network of connected gardens through which amphibians and small mammals, including hedgehogs, can travel. Ensure container ponds have "ladders" in place (see pp.54–55), and add a ramp or large log or stone to steep-sided ponds, which will ensure frogs can enter and exit easily.

Now's the time to prepare for the year ahead by completing any last-minute maintenance.

COMMON FROG
Rana temporaria

Found throughout Europe except for the Mediterranean, the common frog is one of the most likely inhabitants of garden ponds. It eats flying insects, slugs, snails and worms, catching them as they fly or slither past. In summer, you may see frogs resting in your pond, keeping cool while waiting for low-flying insects to fly past.

BREEDING
From late winter males and females emerge from hibernation and travel to their breeding ground, often for a large "party" of hundreds of frogs. Unlike common toads, which are faithful to ancestral breeding grounds, frogs are more likely to try new breeding sites and so they are some of the first amphibians that might turn up in your new pond.

When mating, the male grasps the female in a position known as "amplexus", and fertilizes her eggs as she releases them. Sometimes "mating balls" occur, where several males cling on to, and compete for, a female. This is dangerous for the female, but if she survives she will

Frogs have wet skin, and need to be near water for most of the year, so are more reliant on ponds than other amphibians.

Females are usually slightly larger than the male.

mate with the strongest, most determined male, possibly making the risk of death worth it.

Spawn is laid in clumps, usually en masse with spawn of other frogs. Each female releases one clump. The eggs hatch into tadpoles a couple of weeks later, depending on the weather. They eat pond algae and other detritus, forming a taste for meat as they mature, and in turn are eaten by many other species, including birds, aquatic invertebrates and newts.

Frog tadpoles develop their back legs first, then their front legs, then they absorb their tails and emerge from the pond in midsummer. They hide in long grass and other dense planting, eating insects and small invertebrates before hibernation in autumn. They reach sexual maturity after two to three years.

Frogs hibernate in compost heaps, log piles, or simply buried in the mud. Males sometimes hibernate at the bottom of ponds, presumably so they can be the first to the party in spring.

Female frog
Females develop eggs in summer and hibernate with them, ready for breeding the following spring.

FROGS MATING
As they emerge from overwintering, frogs set off to breed. To mate, the male frog grasps the female around her back in a hold known as "amplexus", and then he fertilizes her eggs as she lays them.

FROG TADPOLES
Tadpoles start off black in colour and mature to gold-speckled brown. They often hide among pond plants for safety as many other pond species prey upon them.

FROGLET
Once the tadpoles have all four legs and have absorbed their tails, they start to venture out of the pond, usually in early summer, after rain. They need a gentle slope so they can clamber out of the pond.

CARING FOR YOUR POND

POND CARE AND MAINTENANCE

As your pond matures, it will change, graduating from a pool of water with some plants in, to an amazingly complex world of multi-layered ecosystems. Plants will grow and die, animals will come and go, pond levels will fluctuate, and even the pond depth will change over time as natural processes slowly fill it.

In the few weeks after digging your pond it will probably turn green. This is perfectly natural as algae forms in the water before the plants start into growth. It's something every one of my ponds has done initially. I would go as far as to say that all new ponds turn green, and then settle as the plants start growing. It's just ponds. Algae is a natural, essential feature of ponds – tadpoles and other species eat it. However some algae can get out of hand and you may need to act (see pp.172–173). But don't panic – use only rain water to top up if necessary, and add plenty of floating, oxygenating and marginal pond plants if you haven't

already. The water should clear in a few weeks and if it doesn't there are measures you can take to control it.

Your pond should need minimal maintenance: removing leaves, topping up the water and clearing snow off ice (see p.174). Other problems can be trickier. Over the years, I've had to remove lots of blanketweed, repair leaks, and even replace a pond liner that foxes chewed holes in! Some common questions about pond care are covered on pages 176–179.

Build up of blanketweed
Sometimes blanketweed can build up. Adding more pond plants and using barley straw can prevent algal blooms.

HOW TO MAINTAIN YOUR POND

All ponds need a bit of maintenance now and again. Anything you do to your pond disturbs wildlife, so keep such disturbances to a minimum during spring and summer, when so many species are using the pond. Autumn is the best time to do bigger tasks, when there are fewer risks to the animals in the water.

REGULAR TASKS

CLEAR AWAY LEAVES AND OTHER DETRITUS

WHEN? *Anytime*

Use a net to gently sweep out leaves and other material from your pond. Do this regularly after heavy leaf fall, remembering to always swill it in a bucket of pond water to check for wildlife that may have come with it. Even in autumn, animals such as water hoglice and dragonfly larvae can be caught out. Gently release these back into the pond before composting the leafy material.

REMOVE BLANKETWEED AND DUCKWEED

WHEN? *Anytime*

Use a long stick to remove blanketweed by twirling it around so that it wraps around the stick like candy floss. For duckweed, use a net or sieve – a kitchen sieve is remarkably efficient – to gently

Removing leaves
A few leaves in your pond won't be a problem, but scoop out excess leaves so they don't build up at the bottom.

Trim pond plants
Trim back pond plants to create some open water on the surface of your pond.

clear the surface. Swill all material you remove in a bucket of pond water before composting, as some animals will hide in it. I always leave blanketweed in a bucket for a couple of days after removing it. That way, I can rescue and return to the water anything that I might have missed the first time around, including dragonfly larvae, baby frogs, toads, tadpoles and smaller invertebrates. Deter blanketweed and duckweed from building up by suppressing the nutrients they need to grow (see p.178).

CUT BACK POND PLANTS
WHEN? *Anytime*
Pond plants should ideally cover around two-thirds of your pond, absorbing nutrients as they grow, and helping to suppress algae. Every now and then you'll need to use secateurs to trim them back, particularly those that "raft" over the surface. This will allow more light into the lower layers of the pond so that plants under the surface can grow strongly.

REPOT POND PLANTS
WHEN? *Spring or early autumn*
Use large gloves to gently lift the plant pots out of the pond; don't worry if the roots have grown beyond the pot into the silt beneath them. Once the plant is out of the water, cut back all stems to a manageable level and trim the roots growing out of the pot. Then, gently remove the plant from the pot. Repot the whole plant into a larger planting basket with fresh aquatic compost. If the rootball is getting very large, or if you want to make new plants, you can divide it (see p.174). Top with gravel or stones to stop the compost washing away, then gently lower the plant back to the same depth as it was previously in the water.

CUT BACK MARGINALS AND EDGE PLANTS
WHEN? *Autumn*
Cut back plants around the edge of the pond that overhang the water, otherwise foliage might fall in and cause problems as it breaks down. Make sure you're in a safe position, preferably kneeling, to avoid accidentally falling into the water.

Stop ice from sealing the pond
Float a ball in your pond so that a
small area remains free of ice.

will also enable plants below to continue
photosynthesizing, providing oxygen for
animals overwintering in the pond.

OCCASIONAL TASKS

Use secateurs to trim back overhanging
foliage, leaving plenty at the edge as
shelter for overwintering insects and
even small amphibians. Trim marginals to
control wayward growth (brooklime, for
example, can travel quite far on land).

PREVENT ICE AND REMOVE SNOW
WHEN? *Winter*
To avoid the rare, potential risk to
wildlife caused by the pond freezing
over for several weeks at a time (see
p.179), you may want to float a ball on
the surface of the water. This will delay
(rather than prevent) ice developing,
helping to provide a drinking spot for
mammals. Depending on the size of your
pond, use a tennis ball or a football – as
long as it floats. If ice has formed, do not
break it: this can shock – and sometimes
kill – wildlife. Instead, place a saucepan of
boiled water on the ice to gently melt a
hole. Removing snow from the surface

DIVIDE POND PLANTS
WHEN? *Every 3–5 years in spring (or after
flowering for spring-flowering plants)*
Gently remove the pot from the pond.
Cut back any foliage so you can see what
you're doing. Remove the plant from
the basket – you may need to cut it out.
Divide the plant (retaining some growing
points on each section) by pulling or
cutting the rootball apart – an old bread
knife works well for tough rootballs.
Replant into fresh aquatic compost and
add a layer of stones or gravel to weigh
the pot down and settle the compost.
Return the pot to the same shelf.

DE-SILT THE POND
HOW OFTEN *Every 5–10 years; less if you
regularly remove leaves from your pond*
WHEN? *Autumn*
This is a messy, disruptive job. First, empty
all plants and water into buckets, taking
care with any wildlife. Scrape off the top
layer of mud from the bottom of the

pond, which will contain overwintering invertebrates, eggs and other pond life, and carefully store this too. Then remove the lower levels of mud and silt, which you can compost or bury in your garden. Put back the top layer of mud and partly refill with rain water. After a couple of days, return the pond water and plants.

PRUNE BACK OVERHANGING BRANCHES
WHEN? *As needed, autumn and winter*
While you should situate your pond away from overhanging trees or shrubs (see p.22), they may grow over the pond, adding shade and increasing the number of leaves that fall into the water. Cut back overhanging branches, but don't overdo it – the tree will be providing shelter for birds and other wildlife.

REPAIR OR REPLACE A POND LINER
HOW OFTEN *Hopefully never*
WHEN? *Autumn*
Occasionally ponds spring a leak due to stones, twigs and the claws of pets and

wildlife. Small leaks can be hard to find. First make sure dry weather isn't causing lower water levels by topping up the pond at night and checking again in the morning. Look for the leak just above the level of the water, removing stones and logs and checking the liner behind them. Once you've identified a hole, drain the pond some more and let the area you're going to fix fully dry out – put plants and animals into buckets if you need to. Use one of the products for mending a small hole, following the instructions. Some emit harmful fumes, so you may need to wear protective gear.

Sometimes pond liners can become brittle and need replacing. Avoid this by buying liner with a long guarantee and protecting it from the sun by covering it with plants and rocks. If you do need to replace the liner, remove all plants and animals, keeping them in buckets and tanks to return to the pond later. The way you built your pond will determine how you take it apart and re-lay the liner, but once you've got back to the bare soil it's no different to lining a new pond – you may even want to make some changes to the shape and depth while you're at it!

Replacing a liner
Laying new pond liner isn't that different to starting a pond from scratch. Choose liner with a long guarantee and keep it protected from the sun as much as possible.

COMMON QUERIES

Got a question about your pond? Chances
are it will be about algae, planting, fish or
dead frogs. Here, I've answered these
commonly asked questions and added a
few more, which you may find helpful.

Q. Should I add fish to my pond?

Goldfish and other types of ornamental
fish make a beautiful addition to ponds.
Indeed, many types of fish can be a
natural part of a pond's ecosystem.
Unfortunately, in small garden ponds,
fish will eat virtually all invertebrate life,
including tadpoles (although fish are less
likely to eat toad tadpoles as these are
slightly poisonous). To have fish you will
also need to install a pump (see p.24), as
ornamental fish produce a lot of waste in
a small area. If you're keen on fish,
consider two ponds – one for the fish and
one for the wildlife. But otherwise, leave
the fish out. There will be plenty to
entertain you that comes naturally.

Q. Should I only use native plants?

A native plant is one that is indigenous to a
given area over a certain time period. In
the British Isles we consider native plants
to be those that have been growing here
since the end of the last Ice Age. As
they've been growing here for a long time,
our wildlife has evolved alongside these
plants, and so natives tend to be
considered better for wildlife. Some
moths and butterflies, for instance, have
native larval food plants (although other
pollinators have been found to visit both
native and non-native varieties).

In ponds, native plants tend to be less
likely to outgrow the pond or threaten
natural waterways in the wild, though

you should choose plants that are suitable for the size of your pond as some natives are also vigorous. To be on the safe side, I've included only British native plants in the plant chapter of this book (see pp.70–87). You may spot some attractive non-natives at the garden centre but do check how vigorous they are before buying them, and make sure they're not considered invasive (see pp.88–89 and p.183). If you find an invasive plant in your pond, remove it at once. It's always easier to remove a seedling than a large established plant. Dispose of it carefully, preferably by burning or hot composting so that it cannot regrow. Never return these plants to the wild as they can clog up rivers, streams and wild ponds.

Q. Do I need to be worried about algae and duckweed?

There are many species of algae, which feed on nutrients in the water. Algae is a natural part of your pond's ecosystem and is a food source for many aquatic species, including tadpoles. However, it is also capable of reproducing at an alarming rate and can quickly become a problem in ponds if you're not careful.

The first algae you are likely to encounter in your pond is the kind that forms the green, soupy look new ponds develop before they become established. This is normal and will usually sort itself out as the pond settles. The other common algae is actually made up of a few species known collectively as "blanketweed". These can clog up ponds, forming dense, hair-like mats near the surface which block light and oxygen from the rest of the pond.

Duckweed is not an algae but is a floating plant that is made up of tiny circular leaves that sit on the surface of the water and reproduce rapidly. Like algae, it's a normal and natural part of the pond's ecosystem; indeed some animals actually rely on duckweed to live (the duckweed weevil, *Tanysphyrus lemnae*, is particularly lovely). Problems start to occur when it covers the whole of the pond, blocking light and oxygen to anything beneath it.

You can stop algae and duckweed building up by removing them regularly by hand (see pp.172–173). You can also buy bespoke solutions to remove algae from the pond, or add barley straw, which helps to discourage algae (both are available in garden centres and pet shops). Simply drop the barley straw into the pond. A longer-term course of action alongside these measures is to keep your pond's nutrient levels down (see p.178). Growing more plants, as well as using aquatic compost or subsoil, will help create the right balance in your pond.

Q. How can I reduce the nutrients in my pond water?

Algae and duckweed thrive in nutrient-rich water, which is why it's important to try to keep nutrient levels as low as possible in your pond. Use nutrient-poor subsoil rather than topsoil when creating a wildlife pond (see pp.32–41), and low-nutrient aquatic compost rather than multipurpose compost or ordinary garden soil in planting baskets. If possible, avoid filling or topping up your pond with tap water, which contains more nutrients than rain water. Growing plenty of pond plants can also help reduce nutrients in the water as they absorb them as they grow. Bear in mind, also, that having fish in a small pond will contribute to algae levels, as any food they don't eat, along with their droppings, will add nutrients.

Q. Should I top up my pond?

Fluctuating water levels are completely normal in ponds, and mimic some naturally occurring ponds that completely dry up in summer. While a dried up pond will result in the death of some species living in it, it's thought that other species might benefit from this. Newts and frogs, for example, may prefer temporary ponds over permanent ones because tadpole predators such as fish

are unable to survive in them. What's more, it's thought frog tadpoles could even speed up their metamorphosis in shallow water (although some will inevitably perish).

With all this being said, however, you will need to ask yourself how much you want to see a dried up pond every time you look out of your window. Bear in mind that exposed pond liner can crack in the sun and you might not feel comfortable seeing tadpoles and other aquatic life struggling in an ever-decreasing puddle.

If you do decide to top up your pond, ideally do so with rain water from your water butt so you don't encourage problems with algae further down the line. If no rain water is available, tap water can be used as a last resort.

Q. Should I remove leaves from my pond?

It's perfectly normal and natural for leaves and other plant material to fall into your pond. In nature, ponds are supposed to be temporary, and leaf fall and other detritus (animal droppings, dead animals, and so on) all gradually fill them up until they eventually mature into a bog garden, before disappearing completely. As gardeners, however, we try to keep our ponds forever young, and that inevitably involves removing the things – leaves, but also algae – that can start to age it.

A few leaves can be good for wildlife: pond snails may lay their eggs on them while tadpoles may eat them and other animals shelter beneath them. However, they also add nutrients to the water (see left), which aids algal growth.

So yes, remove leaves regularly to stop algal blooms and a build up of detritus at the bottom, but perhaps let a few remain for wildlife.

Q. Why are my frogs dying?

Frogs can die for a number of reasons. Many choose to overwinter at the bottom of the pond, breathing through their skin. Sometimes, after a particularly hard winter, you may find dead frogs on the surface after the ice has thawed. It's not known exactly why this happens (and there's much debate about it in scientific circles), but it is thought that a lack of oxygen or a build up of toxic gases under thick ice in ponds with a lot of rotting vegetation at the bottom may play a part.

These deaths are a sad, but natural phenomenon known as "winterkill". Removing leaves in autumn (see opposite) and preventing the pond from icing over completely (see p.174) may help to prevent frog deaths, but there's no hard evidence to confirm this.

You will, in addition, always get a few frog casualties in spring. Breeding is a very energy-intensive process, which often occurs immediately after hibernation – it's amazing that any survive at all. What's more, females can sometimes drown while mating, usually when large mating balls form in which several males compete to mate with her. Again this is completely natural, but no less sad for those of us who find the dead bodies.

Amphibians can also succumb to disease. The main disease affecting garden amphibians is ranavirus, which affects frogs, toads and their tadpoles. Symptoms include reddening of the skin and skin ulcerations. While some frogs die individually, mass death events are common, which can be quite alarming and upsetting. Remove dead frogs immediately to reduce the spread of infection, and avoid moving frogspawn, other animals, plants or even water from one pond to another.

GLOSSARY

Algae A diverse group of aquatic organisms that can photosynthesize. Many inhabit ponds, including several types known as "blanketweed".

Algal bloom A sudden increase in algal growth, visible as coloured water or green matted strands.

Amphibian A cold-blooded vertebrate with aquatic (gill-breathing) larvae and usually terrestrial (lung-breathing) adults. Includes frogs, toads and newts.

Amplexus The mating position of frogs and toads. The male climbs onto the female's back and clasps on to her using his front legs.

Crustacean Aquatic animals in the order Crustacea. In ponds these include water hoglice and water fleas.

Decomposition The breaking down of animal or plant material (organic matter). This is usually aided by bacteria, fungi and invertebrate activity.

Detritus Decaying organic matter (the remains of plant and animal material). Also used to describe eroded objects such as gravel.

Detritivore An organism that feeds on detritus.

Ecosystem A community of plants and animals and their shared environment.

Habitat A space where an organism lives or breeds.

Hibernaculum A habitat used for hibernation. Hibernacula (plural) can be anything from a gap beneath tree bark to the bottom of your pond.

Invasive The spread of (usually) non-native plants, which can threaten the growth of native plants.

Invertebrate An animal without a vertebral column, or backbone.

Larva (plural: larvae) The juvenile form of an invertebrate.

Metamorphosis The transition from juvenile to adult stage. Often takes several stages (instars). Some invertebrates go through "complete metamorphosis", from larva to adult, while others go through "incomplete metamorphosis", from nymph to adult.

Native Plant or animal found naturally in an area, often since a defined time period.

Nitrate Relating to nitrogen and nitric acid, such as minerals required and absorbed by plants to grow.

Non-native Plant or animal that has arrived or been introduced into an area.

Nymph The juvenile stage of an invertebrate, which more closely resembles the adult stage than larvae, and may inhabit the same environment.

Oxygenate To increase oxygen levels to a given space or habitat.

pH The measure of alkalinity or acidity of a given substance, such as soil or water.

Pupa (plural: pupae) The life stage between larva and adult inside which complete metamorphosis takes place.

Silt The layer of mud and detritus that forms at the bottom of ponds; also describes fine grains of soil.

Subsoil The layer of soil beneath the topsoil, which has few nutrients. Often a different colour to topsoil.

Topsoil The upper layer of soil, usually rich in nutrients and plant life.

Vertebrate An animal with an internal backbone.

RESOURCES

BOOKS

Britain's Reptiles and Amphibians by Howard Inns (Princeton University Press, 2011). A fantastic resource for identifying and learning about Britain's reptiles and amphibians.

Freshwater Life: Britain and Northern Europe by M. Greenhalgh and D. Ovenden (Collins, 2007). Need help identifying something you've found in your pond? It's all in here.

Frogs and Toads by Trevor Beebee (Whittet Books, 1985). Out of print now but well worth hunting down in second-hand shops, *Frogs and Toads* gives you loads of insider knowledge on these glorious amphibians.

The Pond Book: A Guide to Management and Creation of Ponds by P. Williams, J. Biggs, M. Whitefield., A. Thorne, S. Bryant, G. Fox and P. Nicolet. (Ponds Conservation Trust, 1999). Excellent information for those digging large ponds. Available from the Freshwater Habitats Trust.

Wildlife Gardening for Everyone and Everything by Kate Bradbury (Bloomsbury, 2019). Even if I say so myself, this is a great all-round book that helps you create habitats for a wide range of wild species in your garden.

The Wildlife Pond Book by Jules Howard (Bloomsbury 2019). Clearly laid out and very detailed, this book is an incredible resource for pond lovers.

ORGANIZATIONS

Amphibian and Reptile Conservation (ARC)
Conservation charity concerned with the protection of amphibians and reptiles.
arc-trust.org

Amphibian and Reptile Groups of the UK (ARG UK)
Collection of local volunteer amphibian and reptile groups, through which you can get involved in local campaigns.
arguk.org

British Dragonfly Society (BDS)
Charity dedicated to the protection of dragonflies and damselflies.
british-dragonflies.org.uk

Freshwater Habitats Trust
Formerly Pondlife, the charity helps protect freshwater habitats. There's lots of expert advice on creating and maintaining freshwater habitats on their website.
freshwaterhabitats.org.uk

Froglife
Involved with the protection of habitats for amphibian and reptiles, Froglife works with schools and local communities, helping to engage people with ponds and the animals that live in them.
froglife.org

Garden Wildlife Health – Zoological Society of London (ZSL)
This project encourages gardeners to observe and record amphibian deaths, helping scientists to monitor and understand amphibian and reptile diseases.
gardenwildlifehealth.org

The Royal Horticultural Society (RHS)
The UK's leading gardening charity, dedicated to improving people's life through plants.
rhs.org.uk/wildlife

The Wildlife Trusts
A collection of 46 local Wildlife Trusts across the UK, helping to preserve local habitats. Discover local events, including training sessions, by visiting your local branch.
wildlifetrusts.org

INVASIVE PLANTS

For a full list of banned pond plants see the RHS page:
rhs.org.uk/advice/profile?pid=429

INDEX

Page numbers in *italics* refer to illustrations

ACKNOWLEDGMENTS

Author acknowledgments:
I would like to thank Ruth, Amy, Jane, Lucy and the rest of the DK team, Helen Bostock and Andy Salisbury from the RHS, and my agent Jane Turnbull. Also a big shout out to the wonderful Jules Howard, for being on hand with pond and plant queries, general reassurance and being such a great friend.

Publisher acknowledgments:
DK would like to thank Nicola Powling and Steve Marsden for providing illustrations, Steve Crozier for image retouching, Tia Sarkar for editorial assistance, Francesco Piscitelli for proofreading, and Vanessa Bird for indexing.

Picture credits:
The publisher would like to thank the following for their kind permission to reproduce their photographs:

(Key: a-above; b-below/bottom; c-centre; f-far; l-left; r-right; t-top)

123RF.com: alekss 146b, Terence Watts 81tl, yuenmingliang 89tr, Nemanja Zotovic 163tr; **Alamy Stock Photo:** AGAMI Photo Agency / Theo Douma 143crb, Arterra Picture Library / Arndt Sven-Erik 147tr, 164br, / Clement Philippe 112bl, 159crb, Andrew Bailey 14, Richard Becker 167crb, Juniors Bildarchiv GmbH 111tr, BIOSPHOTO / Jean-Michel Groult 52bl, blickwinkel / A. Jagel 79clb, / fotototo 152-153, / Hartl 151crb, / Hecker 101crb, 103clb, / Hecker 102tr, Anna Stowe Botanica 32bl, Buiten-Beeld / Jelger Herder 115crb, Buiten-Beeld / Jelger Herder 118-119b, Coatsey 147crb, Denis Crawford 101clb, Derek Croucher 167cr, Ethan Daniels 181, David Tipling Photo Library 95t, DP Wildlife Vertebrates 111crb, DP Wildlife Vertebrates 132bl, Elizabeth Whiting & Associates / ewastock gardens 175bl, FloralImages 84crb, FLPA 93, 138tl, 163crb, 113br, 131crb, 135tr, Tim Gainey 60cl, Tim Graham 120-121, Holly Grogan 156bl, Claire Haskins 150-151c, Robert Henno 87crb, Toby Houlton 163cr, imageBROKER / Arco / G. Lacz 127tr, / Christian Hütter 143tr, Gina Kelly 160cl, Henri Koskinen 83bl, lifes all white 134b, / Michaela Walch 124b, kris Mercer 32cl, mikle15 141tr, Jerome Murray - CC 123b, 148bl, Nature Photographers Ltd / Paul R. Sterry 101cla, 115tr, / Rob Read 108b, Nature Picture Library / 2020VISION / Mark Hamblin 4bl, 122tl, / Kim Taylor 117, / Simon Colmer 111cr, Mike O'Carroll 13b, 182, Papilio / Robert Pickett 109tr, Paul Thompson Images 84tl, Paul Thompson Images / Kathy Traynor 161, Neil Phillips 102crb, 151tr, PjrNature 151cr, Premium Stock Photography GmbH / Frank Teigler 84clb, Gillian Pullinger 113tl, 131tr, 131cra, George Reszeter 157tr, SBP 127cr, Scenics & Science 103crb, Mick Sharp 52cl, Gary K Smith 149tl, Tierfotoagentur / T. Harbig 106tl, Nick Upton 104-105, 107br, Hans Verburg 147cr, Damian Waters 114-115b, Tony Watson 171, Ian West 31, 127crb, WILDLIFE GmbH 145, Rosemary Winnall 135cr **Dreamstime.com:** Aleoks 79tr, Alessandrozocc 89clb, Alexmak72427 75tl, Anitasstudio 110b, Anuraj R V 88crb, Argenlant 76tr, Natalia Bachkova 154tl, Adrian Ciurea 167tr, Creativenaturemedia 75tr, Natalya Erofeeva 80crb, Paul Farnfield 80clb, Jaroslav Frank 136-137, Per Grunditz 143cr, Pawel Horazy 86crb, Javier Alonso Huerta 102clb, Isselee 158b, 162b, Jpldesigns 42cl, Jsmcqueen 102tl, Ferenc Kósa 87tr, Matauw 87clb, Sander Meertins 119crb, Micromann 119tr, Chris Moncrieff 155br, Mps197 85tl, Evgeniy Muhortov 76tl, Oakdalecat 83tl, Martin Pelanek 115cr, Plej92 89crb, Dmitry Potashkin 76bl, Ian Redding 81tr, 89tl, Sarah Richert 165tl, Jonas Rönnbro 159cr, Sandra Standbridge 135crb, 140bc, Stockr 85bl, Wanuttapong Suwannasilp 83tr, Andris Tkacenko 77crb, Pavel Trankov 126, Iva Vagnerova 79crb, Iva Villi 87tl, Whiskybottle 75bl, 79tl, 80tl, 81bl, 128, 77clb, Ian Wilson 9, Rudmer Zwerver 142b, 159tr, 166b **GAP Photos:** Richard Bloom 5b, 32-33b, 42-43bc, Jonathan Buckley / Design: John Massey 10-11t, Carole Drake 60-61b, John Glover / Design Roger Platt - Chelsea FS 1996 52-53bc, Howard Rice / Design: Joy Martin 2, Nicola Stocken 60bl **Getty Images / iStock:** E+ / kevinjeon00 4br, 71, / NNehring 103cla, Jaimie Tuchman 85tr

All other images © Dorling Kindersley
For further information see: www.dkimages.com

ABOUT THE AUTHOR

Kate Bradbury is an award-winning author and journalist, specializing in wildlife gardening. She edits the wildlife pages of BBC *Gardeners' World Magazine*, and regularly writes articles for *The Telegraph*, *The Guardian*, RHS magazine *The Garden*, and The Wildlife Trusts magazines. She is the author of *Wildlife Gardening: For Everyone and Everything* and memoir *The Bumblebee Flies Anyway* (Bloomsbury), and *The Wildlife Gardener* (White Owl). She lives in Brighton, where she created her wildlife garden as part of the BBC Springwatch campaign. She and her garden have also appeared on *BBC Gardeners' World*. She's a patron of Froglife and Bumblebee Conservation Trust, and garden ambassador of Butterfly Conservation.

DK LONDON

Project Editor Amy Slack
Editor Jane Simmonds
Art Editor Nigel Wright
Production Editor David Almond
Production Controller Rebecca Parton
Jacket Designer Nicola Powling
Jacket Editor Lucy Philpott
Managing Editor Ruth O'Rourke
Managing Art Editor Christine Keilty
Consultant Gardening Publisher Chris Young
Art Director Maxine Pedliham
Publishing Directors Mary-Clare Jerram, Katie Cowan

ROYAL HORTICULTURAL SOCIETY

Consultant Helen Bostock
Publisher Rae Spencer-Jones

First published in Great Britain in 2021 by
Dorling Kindersley Limited
DK, One Embassy Gardens, 8 Viaduct Gardens,
London, SW11 7BW

The authorised representative in the EEA is Dorling Kindersley
Verlag GmbH.
Arnulfstr. 124, 80636 Munich, Germany

A CIP catalogue record for this book
is available from the British Library.
ISBN: 978-0-2414-7292-7

Printed and bound in China

For the curious
www.dk.com